Physics of Society: Econophysics and Sociophysics

# Limit Order Books

Frédéric Abergel
Marouane Anane
Anirban Chakraborti
Aymen Jedidi
Ioane Muni Toke

# CAMBRIDGE
UNIVERSITY PRESS

4843/24, 2nd Floor, Ansari Road, Daryaganj, Delhi - 110002, India

Cambridge University Press is part of the University of Cambridge.

It furthers the University's mission by disseminating knowledge in the pursuit of education, learning and research at the highest international levels of excellence.

www.cambridge.org
Information on this title: www.cambridge.org/9781107163980

© Authors 2016

This publication is in copyright. Subject to statutory exception and to the provisions of relevant collective licensing agreements, no reproduction of any part may take place without the written permission of Cambridge University Press.

First published 2016

Printed in India by Shree Maitrey Printech Pvt. Ltd., Noida

*A catalogue record for this publication is available from the British Library*

ISBN 978-1-107-16398-0 Hardback

Cambridge University Press has no responsibility for the persistence or accuracy of URLs for external or third-party internet websites referred to in this publication, and does not guarantee that any content on such websites is, or will remain, accurate or appropriate.

# Contents

| | |
|---|---|
| *Figures* | xi |
| *Tables* | xv |
| *Foreword* | xvii |
| *Preface* | xix |
| *Acknowledgments* | xxi |

**1 Introduction**    **1**

## PART ONE
## EMPIRICAL PROPERTIES OF ORDER-DRIVEN MARKETS

**2 Statistical Properties of Limit Order Books: A Survey**    **9**

   2.1 Introduction    9
   2.2 Time of Arrivals of Orders    9
   2.3 Volume of Orders    12
   2.4 Placement of Orders    13
   2.5 Cancellation of Orders    15
   2.6 Average Shape of the Order Book    16
   2.7 Intraday Seasonality    18
   2.8 Conclusion    19

**3 The Order Book Shape as a Function of the Order Size**    **20**

   3.1 Introduction    20
   3.2 Methodology    20
   3.3 The Regression Model    22
   3.4 Conclusion    28

**4 Empirical Evidence of Market Making and Taking**    **29**

   4.1 Introduction    29
   4.2 Re-introducing Physical Time    29
   4.3 Dependency Properties of Inter-arrival Times    31

|  |  |
|---|---|
| 4.3.1 Empirical evidence of market making | 31 |
| 4.3.2 A reciprocal effect? | 33 |
| 4.4 Further Insight into the Dependency Structure | 35 |
| 4.4.1 The fine structure of inter-event durations: Using lagged correlation matrices | 37 |
| 4.5 Conclusion | 41 |

## PART TWO
## MATHEMATICAL MODELLING OF LIMIT ORDER BOOKS

| | |
|---|---|
| **5 Agent-based Modelling of Limit Order Books: A Survey** | **45** |
| 5.1 Introduction | 45 |
| 5.2 Early Order-driven Market Modelling: Market Microstructure and Policy Issues | 47 |
| 5.2.1 A pioneer order book model | 47 |
| 5.2.2 Microstructure of the double auction | 48 |
| 5.2.3 Zero-intelligence | 48 |
| 5.3 Order-driven Market Modelling in Econophysics | 49 |
| 5.3.1 The order book as a reaction-diffusion model | 49 |
| 5.3.2 Introducing market orders | 51 |
| 5.3.3 The order book as a deposition-evaporation process | 53 |
| 5.4 Empirical Zero-intelligence Models | 54 |
| 5.5 Some Analytical and Mathematical Developments in Zero-intelligence Order Book Modelling | 57 |
| 5.6 Conclusion | 58 |
| **6 The Mathematical Structure of Zero-intelligence Models** | **59** |
| 6.1 Introduction | 59 |
| 6.1.1 An elementary approximation: Perfect market making | 59 |
| 6.2 Order Book Dynamics | 61 |
| 6.2.1 Model setup: Poissonian arrivals, reference frame and boundary conditions | 61 |
| 6.2.2 Evolution of the order book | 64 |
| 6.2.3 Infinitesimal generator | 66 |
| 6.2.4 Price dynamics | 67 |
| 6.3 Ergodicity and Diffusive Limit | 68 |
| 6.3.1 Ergodicity of the order book | 69 |
| 6.3.2 Large-scale limit of the price process | 70 |

|       |       |                                                                                  |     |
| ----- | ----- | -------------------------------------------------------------------------------- | --- |
|       | 6.3.3 | Interpreting the asymptotic volatility                                           | 74  |
| 6.4   |       | The Role of Cancellations                                                        | 75  |
| 6.5   |       | Conclusion                                                                       | 76  |

## 7 The Order Book as a Queueing System  77

- 7.1 Introduction  77
- 7.2 A Link Between the Flows of Orders and the Shape of an Order Book  78
  - 7.2.1 The basic one-sided queueing system  78
  - 7.2.2 A continuous extension of the basic model  80
- 7.3 Comparison to Existing Results on the Shape of the Order Book  83
  - 7.3.1 Numerically simulated shape in Smith et al. (2003)  83
  - 7.3.2 Empirical and analytical shape in Bouchaud et al. (2002)  84
- 7.4 A Model with Varying Sizes of Limit Orders  88
- 7.5 Influence of the Size of Limit Orders on the Shape of the Order Book  92
- 7.6 Conclusion  96

## 8 Advanced Modelling of Limit Order Books  97

- 8.1 Introduction  97
- 8.2 Towards Non-trivial Behaviours: Modelling Market Interactions  97
  - 8.2.1 Herding behaviour  98
  - 8.2.2 Fundamentalists and trend followers  99
  - 8.2.3 Threshold behaviour  101
  - 8.2.4 Enhancing zero-intelligence models  101
- 8.3 Limit Order Book Driven by Hawkes Processes  102
  - 8.3.1 Hawkes processes  103
  - 8.3.2 Model setup  104
  - 8.3.3 The infinitesimal generator  105
  - 8.3.4 Stability of the order book  106
  - 8.3.5 Large scale limit of the price process  108
- 8.4 Conclusion  110

## PART THREE
## SIMULATION OF LIMIT ORDER BOOKS

## 9 Numerical Simulation of Limit Order Books  113

- 9.1 Introduction  113
- 9.2 Zero-intelligence Limit Order Book Simulator  113
  - 9.2.1 An algorithm for Poissonian order flows  113
  - 9.2.2 Parameter estimation  115

|  |  |
|---|---|
| 9.2.3 Performances of the simulation | 120 |
| 9.2.4 Anomalous diffusion at short time scales | 127 |
| 9.2.5 Results for CAC 40 stocks | 129 |
| 9.3 Simulation of a Limit Order Book Modelled by Hawkes Processes | 129 |
| 9.3.1 Simulation of the limit order book in a simple Hawkes model | 130 |
| 9.3.2 Algorithm for the simulation of a Hawkes process | 130 |
| 9.3.3 Parameter estimation | 133 |
| 9.3.4 Performances of the simulation | 134 |
| 9.4 Market Making and Taking, Viewed from a Hawkes-process Perspective | 138 |
| 9.5 Conclusion | 141 |

## PART FOUR
## IMPERFECTION AND PREDICTABILITY IN ORDER-DRIVEN MARKETS

| | |
|---|---|
| **10 Market Imperfection and Predictability** | **145** |
| 10.1 Introduction | 145 |
| 10.2 Objectives, Methodology and Performances Measures | 146 |
| 10.2.1 Objectives | 146 |
| 10.2.2 Methodology | 147 |
| 10.2.3 Performance measures | 148 |
| 10.3 Conditional Probability Matrices | 148 |
| 10.3.1 Binary case | 150 |
| 10.3.2 Four-class case | 154 |
| 10.4 Linear Regression | 156 |
| 10.4.1 Ordinary least squares (OLS) | 156 |
| 10.4.2 Ridge regression | 158 |
| 10.4.3 Least Absolute Shrinkage and Selection Operator (LASSO) | 163 |
| 10.4.4 Elastic net (EN) | 165 |
| 10.5 Conclusion | 167 |
| *Appendix A* **A Catalogue of Order Types** | 169 |
| *Appendix B* **Limit Order Book Data** | 171 |
| *Appendix C* **Some Useful Mathematical Notions** | 176 |
| *Appendix D* **Comparison of Various Prediction Methods** | 187 |
| *Bibliography* | 209 |

# Figures

1.1 A schematic illustration of the order book.   3
2.1 Distribution of interarrival times for stock BNPP.PA in log-scale.   10
2.2 Distribution of interarrival times for stock BNPP.PA.   10
2.3 Distribution of the number of trades in a given time period $\tau$ for stock BNPP.PA.   11
2.4 Distribution of volumes of market orders.   12
2.5 Distribution of normalized volumes of limit orders.   13
2.6 Placement of limit orders.   14
2.7 Placement of limit orders.   14
2.8 Distribution of estimated lifetime of cancelled limit orders.   15
2.9 Distribution of estimated lifetime of executed limit orders.   16
2.10 Average quantity offered in the limit order book.   17
2.11 Average limit order book: price and depth.   17
2.12 Normalized average number of market orders in a 5-minute interval.   18
2.13 Normalized average number of limit orders in a 5-minute interval.   19
3.1 Mean-scaled shapes of the cumulative order book.   22
3.2 Mean-scaled number of market orders.   24
3.3 Mean-scaled shapes of the cumulative order book.   25
4.1 Empirical distribution function of the bid-ask spread in event time and in physical time.   30
4.2 Empirical distributions of the time intervals between two consecutive orders and of the time intervals between a market order and the immediately following limit order.   32
4.3 Empirical distributions of the time intervals between a market order and the immediately following limit order.   33

4.4 Empirical distributions of the time intervals between two consecutive orders and of the time intervals between a limit order and an immediately following market order. 34
4.5 Impact functions on $M^1_{buy}$ arrival intensity. 39
4.6 Impact functions on the six events $O^1$. 40
5.1 Illustration of the Bak, Paczuski and Shubik model. 50
5.2 Snapshot of the limit order book in the Bak, Paczuski and Shubik model. 51
5.3 Empirical probability density functions of the price increments in the Maslov model. 52
5.4 Average return $\langle r_{\Delta t} \rangle$ as a function of $\Delta t$ for different sets of parameters and simultaneous depositions allowed in the Challet and Stinchcombe model. 54
5.5 Lifetime of orders for simulated data in the Mike and Farmer model. 56
5.6 Cumulative distribution of returns in the Mike and Farmer model. 56
6.1 Order book dynamics. 63
7.1 Shape (top panel) and cumulative shape (bottom panel) of the order book. 85
7.2 Comparison of the shapes of the order book in our model (black curves) and using the formula proposed by Bouchaud et al. (2002) (gray curves). 86
7.3 Price density function $\pi_{pA}$ as a function of the price. 88
7.4 Shape of the order book as computed in Eq. (7.34) (top) and cumulative shape of the order book as computed in Eq. (7.33) (bottom). 93
7.5 Shape of the order book as computed in Eq. (7.34). 94
7.6 Shape of the order book as computed in Eq. (7.34). 95
9.1 Model parameters: arrival rates and average depth profile (parameters as in Table 9.2). 118
9.2 Model parameters: volume distribution. Panels $(a)$, $(b)$ and $(c)$ correspond respectively to market, limit and cancellation orders volumes. 119
9.3 Average depth profile. 120
9.4 Probability distribution of the spread. 121
9.5 Autocorrelation of price increments. 122
9.6 Price sample path. 122
9.7 Probability distribution of price increments. 123
9.8 Q-Q plot of mid-price increments. 123
9.9 Signature plot: $\sigma_h^2 := \mathbb{V}\left[P(t+h) - P(t)\right]/h$. 125
9.10 A cross-sectional comparison of liquidity and price diffusion characteristics between the model and data for CAC 40 stocks (March 2011). 126

| | | |
|---|---|---|
| 9.11 | Simulation of a two-dimensional Hawkes process with parameters given in Eq. (9.8) | 132 |
| 9.12 | Simulation of a two-dimensional Hawkes process parameters given in Eq. (9.8). (Zoom of Fig. 9.11) | 132 |
| 9.13 | Empirical density function of the distribution of the durations of market orders (left) and limit orders (right) for three simulations. | 135 |
| 9.14 | Empirical density function of the distribution of the time intervals between a market order and the following limit order for three simulations. | 136 |
| 9.15 | Empirical density function of the distribution of the bid-ask spread for three simulations. | 136 |
| 9.16 | Empirical density function of the distribution of the bid-ask spread for three simulations. | 137 |
| 9.17 | Empirical density function of the distribution of the 30-second variations of the mid-price for five simulations. | 138 |
| 9.18 | Hawkes parameters for aggressive limit orders for various CAC40 stocks. | 139 |
| 9.19 | Hawkes parameters for aggressive market orders for various CAC40 stocks. | 140 |
| 10.1 | The quality of the binary prediction: The AUC and the Accuracy are higher than 50%. | 151 |
| 10.2 | The quality of the binary prediction. | 151 |
| 10.3 | The quality of the binary prediction. | 152 |
| 10.4 | The quality of the binary prediction. | 152 |
| 10.5 | The quality of the binary prediction. | 153 |
| 10.6 | The quality of the 4-class prediction. | 155 |
| 10.7 | The quality of the OLS prediction. | 157 |
| 10.8 | The quality of the OLS prediction. | 158 |
| 10.9 | The quality of the OLS prediction. | 159 |
| 10.10 | The quality of the Ridge HKB prediction. | 161 |
| 10.11 | The quality of the Ridge LW prediction. | 162 |
| 10.12 | The quality of the Ridge prediction. | 162 |
| 10.13 | The quality of the LASSO prediction. | 163 |
| 10.14 | The quality of the LASSO prediction. | 164 |
| 10.15 | The quality of the LASSO prediction. | 164 |
| 10.16 | The quality of the EN prediction. | 165 |
| 10.17 | The quality of the EN prediction. | 166 |

# Tables

3.1 Basic statistics on the number of orders and the average volumes of orders per 30-minute time interval for each stock. ... 21

3.2 Panel regression results for the models defined in Eq. (3.1), using raw data, for $B_{k,t} = B_{k,t}^5$ (top panel), $B_{k,t} = B_{k,t}^{10}$ (middle panel) and $B_{k,t} = B_{k,t}^{10} - B_{k,t}^5$ (lower panel). ... 26

3.3 Panel regression results for the models defined in Eq. (3.1), using deseasonalized data, for $B_{k,t} = B_{k,t}^5$ (top panel), $B_{k,t} = B_{k,t}^{10}$ (middle panel) and $B_{k,t} = B_{k,t}^{10} - B_{k,t}^5$ (lower panel). ... 27

4.1 Event types definitions. ... 35

4.2 Event occurrences statistics summary. ... 36

4.3 Conditional probabilities of occurrences per event type. ... 36

4.4 Median conditional waiting time. ... 38

4.5 Mean conditional waiting time. ... 38

5.1 Analogy between the $A + B \to \emptyset$ reaction model and the order book in Bak et al. (1997) ... 50

5.2 Analogy between the deposition-evaporation process and the order book in Challet and Stinchcombe (2001). ... 53

5.3 Results of Smith et al. ... 58

9.1 Model parameters for the stock SCHN.PA (Schneider Electric) in March 2011 (23 trading days). ... 116

9.2 Model parameters for the stock SCHN.PA (Schneider Electric) in March 2011 (23 trading days). ... 116

9.3 CAC 40 stocks regression results. ... 129

9.4 Estimated values of parameters used for simulations. ... 133

10.1 Historical occurrences matrix for Deutsche Telekom over 2013. ... 148

10.2 Monthly historical conditional probabilities. ... 149

B.1 Tick by tick data file sample. ... 172

B.2 Trades data file sample. ... 173

B.3 Number of limit and markets orders. ... 174

D.1 The quality of the binary prediction: 1-minute prediction AUC and accuracy per stock. — 187

D.2 The quality of the binary prediction: The daily gain average and standard deviation for the 1-minute prediction (without trading costs). — 189

D.3 The quality of the binary prediction: The daily gain average and standard deviation for the 1-minute prediction (with trading costs). — 190

D.4 The quality of the 4-class prediction: 1-minute prediction AUC and accuracy per stock. — 192

D.5 The quality of the 4-class prediction: The daily gain average and standard deviation for the 1-minute prediction (without trading costs). — 194

D.6 The quality of the 4-class prediction: The daily gain average and standard deviation for the 1-minute prediction (with trading costs). — 195

D.7 The quality of the 4-class prediction: The daily gain average and standard deviation for the 30-minute prediction (without trading costs). — 197

D.8 The quality of the binary prediction: The daily gain average and standard deviation for the 30-minute prediction (with 0.5 bp trading costs). — 199

D.9 The quality of the OLS prediction: The AUC and the accuracy per stock for the different horizons. — 200

D.10 The quality of the Ridge HKB prediction: The AUC and the accuracy per stock for the different horizons. — 202

D.11 The quality of the Ridge LW prediction: The AUC and the accuracy per stock for the different horizons. — 204

D.12 The quality of the LASSO prediction: The AUC and the accuracy per stock for the different horizons. — 205

# Foreword

When physicists became convinced that matter was not continuous but made from atoms, new ideas on old subjects started flourishing. Not only well-known macroscopic laws (thermodynamics, hydrodynamics) became better understood and bolstered by a more fundamental underlying reality, but a host of spectacular and often unexpected effects were rationalized, in particular collective emergent phenomena phase transitions, superconductivity, avalanches, etc. Similarly, after decades of mathematical finance devoted to the study of effective low frequency models of markets (chiefly based on variations on the Brownian motion), the increasing availability of high frequency data now allows a comprehensive study of price formation and of the microstructure of supply and demand. A new era of financial modelling is opening up, with the hope of addressing a hitherto neglected yet crucial aspect of price dynamics: feedback effects that can lead to market anomalies, instabilities and crashes. Instead of considering the market as an inert, reliable measurement apparatus that merely reveals the fundamental value of assets without influencing it, the empirical study of the order book reveals that markets do generate their own dynamics. New intuitions about market dynamics are necessary. New fascinating statistical regularities are collected and modelled, in particular using numerical simulations of agent based models. New analytical tools are being built to account for these observations. The final goal is, much as in physics, to understand the emergent phenomena and replace ad-hoc models of prices by micro-founded ones where jumps, fat-tails and clustered volatility would have a clear origin. This is important on many counts: while the intellectual endeavour is of course exciting in itself, its offshoots will deeply influence the way we think about market regulation in the wake of high-frequency trading, and the models we use for financial engineering (from derivative pricing to algorithmic trading and optimal execution).

**Limit order books** offers a much needed, broad review of a field that has literally exploded in the last 20 years, where researchers from economics, financial mathematics, physics, computer science, etc. compete and confront. This diversity is well illustrated by the content of the present book that covers a very wide ground, from empirical facts to advanced mathematical techniques and numerical simulation tools. It will be a very useful and inspiring entry point for all scientists, engineers, regulators and traders interested in understanding how financial markets really work at the basic level.

Jean-Philippe Bouchaud
Capital Fund Management & École Polytechnique

# Preface

The Chair of Quantitative Finance was created at École Centrale Paris, now CentraleSupélec, in 2007. Since its inception, most of its research activities were devoted to the study of high frequency financial data. The interdisciplinary nature of the team, composed of mathematicians, financial engineers, computer scientists and physicists, gave it a special dimension. A sizeable portion of its research efforts has been focused on the characterization and mathematical modelling of *limit order books*.

Literally at the core of every modern, electronic financial market, the limit order book has triggered a huge amount of research in the past twenty years, marked by the seminal work of Biais et al. (1995) on the empirical analysis of the Paris exchange and revitalized a few years later, in a fascinating manner, by the work of Smith et al. (2003). However, much as this topic is interesting, important and challenging, we realized that there was still no reference book on the subject! We therefore decided to assemble in a single document a survey of the existing literature and our own contributions on limit order books, whether they were pertaining to their statistical properties, mathematical modelling or numerical simulation.

We have tried to follow the intellectual approach of an experimental physicist: empirical data should come first, and only empirical analyses may be considered as a reliable ground for building up any kind of theory. The mathematical modelling follows. Models address the different phenomena that are observed and highlighted, and provide a framework to explain and reproduce these phenomena, and they are studied from theoretical, analytical and numerical perspectives.

The book is thus organized as follows: The first part is devoted to the empirical properties of limit order books; the second part, to their mathematical modelling and the third, to their numerical analysis. The fourth part deals with some advanced topics such as imperfection and predictability. Each part presents a survey of the existing scientific literature, as well as our own contributions.

Significant parts of the material covered in this book have already been presented in bits and pieces in different research and survey articles, in particular Chakraborti et al. (2011a,b); Abergel and Jedidi (2013, 2015); Anane and Abergel (2015); Muni Toke (2015, 2011). However, what was lacking was a consistent and systematic compilation of these, found in a single place where the emphasis was set on a single object of interest. We hope that this book will fulfil this need and complement the already existing abundant literature on market microstructure. The interdisciplinary approaches, with the stress on both empirical data analyses and theoretical studies, will hopefully render it useful to the reader – researcher, graduate student or practitioner, while facilitating him/her in finding most of the contemporary knowledge on this essential component of financial markets.

# Acknowledgments

We are grateful to all our collaborators: Nicolas Huth, Anton Kolotaev, Mehdi Lallouache, Nicolas Millot, Marco Patriarca, Fabrizio Pomponio, Mauro Politi, Rémi Tachet, Riadh Zaatour and Ban Zheng, for their contributions to these developments. We also acknowledge Damien Challet, Rémy Chicheportiche, Charles-Albert Lehalle, Grégoire Loeper, Eric Moulines, François Roueff, Mathieu Rosenbaum, Stéphane Tyc, Nakahiro Yoshida for fruitful discussions and inputs.

We are thankful to BNP Paribas for their generous funding to the Chair of Quantitative Finance.

Some sections of this book were written while Ioane Muni Toke was a Senior Fellow in the "Broad Perspectives and New Directions in Financial Mathematics" program of the Institute for Pure and Applied Mathematics, University of California at Los Angeles.

Ioane Muni Toke also acknowledges the support of the CREST project of the Japan Science and Technology Agency.

Several sections of this book were completed while Frédéric Abergel was visiting the Graduate School of Mathematical Sciences at the University of Tokyo, and the Laboratoire de Probabilités et Modèles Aléatoires at CNRS, Université Pierre et Marie Curie and Université Denis Diderot. He is grateful for the support of these institutions.

Some of the material presented in this book has been previously published and is used here with kind permission from Springer Science+Business Media, Taylor and Francis and World Scientific Publishing:

- Econophysics of Order-driven Markets, "Market Making in an Order Book Model and Its Impact on the Spread", pp. 49–64, Ioane Muni Toke. Copyright Springer-Verlag Italia 2011.
- "Econophysics review: I. Empirical facts", Anirban Chakraborti, Ioane Muni Toke, Marco Patriarca and Frédéric Abergel, Quantitative Finance, 11: 7, pp. 991–1012. Copyright Taylor and Francis 2011. http://www.tandfonline.com
- "Econophysics review: II. Agent-based models", Anirban Chakraborti, Ioane Muni Toke, Marco Patriarca and Frédéric Abergel, Quantitative Finance, 11: 7, pp. 1013–1041. Copyright Taylor and Francis 2011. http://www.tandfonline.com

"A Mathematical Approach to Order Book Modelling", Frédéric Abergel and Aymen Jedidi, International Journal of Theoretical and Applied Finance, 16. Copyright World Scientific Publishing 2013.

Econophysics and Data Driven Modelling of Market Dynamics, "Empirical Evidence of Market Inefficiency: Predicting Single-Stock Returns", pp. 3–66, Marouane Anane and Frédéric Abergel. Copyright Springer International Publishing 2015.

"The order book as a queueing system: average depth and influence of the size of limit orders", Ioane Muni Toke, Quantitative Finance, 15: 5, pp. 795–808. Copyright Taylor and Francis 2015. http://www.tandfonline.com

# CHAPTER 1

# Introduction

"One of the funny things about the stock market is that every time one person buys, another sells, and both think they are astute." – *William Feather, American publisher and author (1889–1981)*

**What is a limit order book**? It is a device that the vast majority of organized electronic markets (all equity, futures and other listed derivatives markets) use to store in their central computer the list of all the interests of market participants. It is essentially a file in a computer that contains *all* the *orders* sent to the market, with their characteristics such as the **sign** of the order (buy or sell), the **price**, the **quantity**, a **timestamp** giving the time the order was recorded by the market, and a host of various market-dependent information. In other words, the limit order book contains, at any given point in time, on a given market, the list of all the transactions that one could possibly perform on this market. Its evolution over time describes the way the market moves under the influence of its participants. In fact, the study of limit order books can provide deep insight into the understanding of the financial market, which is an excellent example of an evolving "complex system" where the different participants *collectively interact* to find the best price of an asset. Hence, this field attracts mathematicians, economists, statistical physicists, computer scientists, financial engineers, and many others, besides the practitioners.

A market in which buyers and sellers meet *via* a limit order book is called an **order-driven market**. In order-driven markets, buy and sell orders are matched as they arrive over time, subject to some priority rules. Priority is always based on price, and then, in most markets, on time, according to a *FIFO* (First In, First Out) rule. Such priority rules are enforced in the vast majority of financial markets, although there exist some notable exceptions or variants: For instance, the Chicago Mercantile Exchange (CME) uses for some order books a *prorata* rule in place of (or together with) time priority. Several different market mechanisms have been studied in the microstructure literature, see for example, Garman (1976); Kyle (1985); Glosten (1994); O'Hara (1997); Biais et al. (1997); Hasbrouck (2007). We will not review these mechanisms in this book [except Garman (1976) in Chapter 5], and rather keep our focus on the almost universal standard of **price/time** priority.

Essentially, three types of orders can be submitted:

- *Limit order* An order to specify a price at which one is willing to buy or sell a certain number of shares, with their corresponding price and quantity, at any point in time;
- *Market order* An order to *immediately* buy or sell a certain quantity, at the best available opposite quote;
- *Cancellation order* An order to cancel an existing limit order.

In the literature dealing with limit order books or market microstructure, agents who submit limit orders are referred to as *liquidity providers*, while those who submit market orders are referred to as *liquidity takers*. In real markets, ever since the various deregulation waves hit the US markets in 2005 and the European markets in 2007 [see for instance, Abergel et al. (2014)], there is no such thing as a *pure* liquidity provider or taker, and this classification should be understood as a convenient shorthand rather than a realistic description of the behaviour of market participants.

It is to be noted that depending on the market under consideration, there exist many variations of the three basic types of orders described above. A catalogue of real order types one can encounter on financial markets is given in Appendix A. Needless to say, the list provided is not exhaustive, and will be expanding over time. In this book, for practical reasons related to the structure of the available datasets, and because we are mainly interested in understanding and modelling universal features of limit order books, the focus will be on a somewhat stylized view of the market where orders can be simply of the "market", "limit" or "cancellation" type.

Limit orders are stored in the order book, until they are either executed against an incoming market order or canceled. The *ask* price $P^A$ (or simply the ask) is the price of the best (i.e. lowest) limit sell order. The *bid* price $P^B$ (or simply the bid) is the price of the best (i.e. highest) limit buy order. The gap between the bid and the ask

$$S := P^A - P^B, \qquad (1.1)$$

is always positive, and is called the *spread*. We define the *mid-price* as the average between the bid and the ask

$$P := \frac{P^A + P^B}{2}. \qquad (1.2)$$

Prices are not continuous, but rather have a discrete resolution $\Delta P$, the *tick size*, which represents the smallest quantity by which they can change.

**Why study limit order books?** It is clear that the study of the empirical properties, as well as the mathematical modelling and numerical simulation, of limit order books, is of paramount importance for the researcher keen on gaining a deep understanding of financial markets.

Traditionally in financial econometrics, the data consist in time series of prices of one or several assets, and models are based on the statistical properties of the various quantities one can build from these time series: Returns, volatility, correlation... However, in order-driven markets, the price dynamics is controlled by the interplay between the incoming order flow and the order book (Bouchaud et al. 2002). Figure 1.1 is a schematic illustration of this process, with the conventional representation of quantities on the bid side by non-positive numbers.

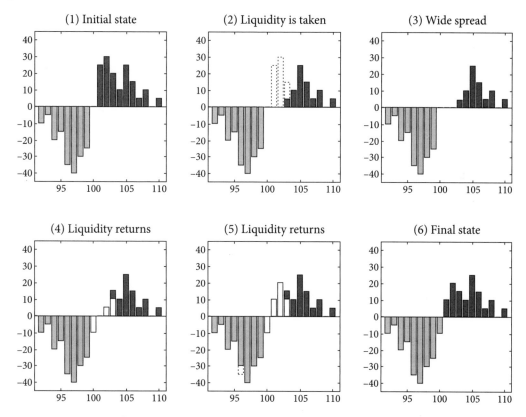

**Fig. 1.1** A schematic illustration of the order book. A buy market order arrives and removes liquidity from the ask side, then, sell limit orders are submitted and liquidity is restored

The study of the limit order book therefore reveals, as a by-product, the price dynamics. One of our main motivations has been to understand the extent to which the mechanisms of the order book have an impact on the price dynamics at the *microstructure* level, and whether this impact remains visible at lower frequencies, i.e. when observing hourly or daily prices. Furthermore, the genuine scientific curiosity for this area of research has recently been very definitely enhanced by the rapid growth of *algorithmic trading* and *high frequency trading*. Market making strategies, optimal execution strategies, statistical arbitrage strategies, being executed at the individual order level, all require a perfect understanding of the limit order book. Some of the statistical properties presented in this book, in particular those pertaining to market imperfections, may be seen as building blocks of such an understanding.

**How to model limit order books?** There are several steps to take when modelling limit order books. Probably, the first one is to select a mechanistic description of the way incoming orders are stored and market orders are executed. This prerequisite is achieved, at least in a stylized form, in all the mathematical models of limit order books, and plays an important role in the simulation of limit order books, for which realistic *matching engines* must be developed in order to study trading strategies. The second step, at a more conceptual and scientifically more fundamental stage, involves choosing a mechanism for the arrival of orders, that is, for the submission of an order of a particular type at a specific date and time. Regarding this aspect, two main approaches have been successful in capturing key properties of the order book—at least to some extent. The first one, led by economists, models the interactions between *rational* agents who act strategically: The agents choose their trading decisions as solutions to individual utility maximization problems [see e.g., Parlour and Seppi (2008), and references therein]. In the second approach, proposed by econophysicists[1], agents are described statistically. In the simplest form along this line of research, the agents are supposed to act *randomly*. This approach is sometimes referred to as *zero-intelligence* order book modeling, in the sense that the arrival times and placements of orders of various types are *random and independent*, the focus being primarily on the "mechanistic" aspects of the continuous double auction rather than the strategic interactions between agents. Despite this apparently unrealistic simplification, statistical models of the order book do capture many salient features of real markets, and exhibit interesting, non-trivial mathematical properties that form the basis of a thorough understanding of limit order books. It is however necessary to depart from this overly simplified approach and study models were agents do interact, at least in a statistical way. Although, there exists a rather vast, fascinating literature on models of financial markets with interacting or competing agents (see e.g., Brock and Hommes

---

[1] Scholars who work in the interdisciplinary field of "Econophysics", comprised of the two fields economics and physics, using ideas and tools from both areas to study complex socio-economic systems.

(1998) Lux and Marchesi (2000)), very little is concerned with order-driven markets. Some recent results in this direction, based on a statistical approach using mutually exciting arrival processes, are presented in this book.

**What is in this book?** Our approach has been to start with the limit order book data, trying to assess their statistical properties. Hence, Chapter 2 is a survey of *stylized facts* on limit order books, Chapter 3 focuses on the shape of the order book and its relation to the size of incoming orders, whereas Chapter 4 is concerned with experimental evidence of the interaction between liquidity providers and takers on order-driven markets. We then moved on to the mathematical models: Chapter 5 is a survey of early works on limit order book modelling; Chapters 6 and 7 present an in-depth, rigourous mathematical theory of zero-intelligence models. In Chapter 8, we review some more advanced agent-based models, and present recent results on limit order books driven by interacting and competing statistical agents. We then provide in Chapter 9 a framework for simulations, and analyze and discuss some numerical results. Finally, in Chapter 10, we return to empirical studies, but with a different, more practical motivation, that of the profitability of trading strategies in order-driven markets.

**What is not in this book?** Obviously, so many things.

For the sake of consistency, we have deliberately left out several alternative approaches to order-driven markets modelling. Whether one actually requires to understand the motivation of the agents in order to obtain a faithful description of their behaviour is an open debate, and we are happy to participate in it with our systematic statistical approach.

Also of great importance is the study of *market impact*. This subject is definitely an important topic, with great practical implications, and although limit order book models obviously offer various possible mechanisms for market impact, we do not address this specific question.

Also connected to market impact, the systematic study of *trading strategies* in order-driven markets in only touched upon in the fourth part of this book, and should be studied at greater lengths.

We could keep on extending this list of regrets. It is clear that progresses must be made in the study of limit order books. Some are already in the making, and we certainly hope that this book will lend an impetus to many others.

even
# PART ONE
# EMPIRICAL PROPERTIES OF ORDER-DRIVEN MARKETS

# CHAPTER 2

# Statistical Properties of Limit Order Books: A Survey

## 2.1 Introduction

The computerization of financial markets in the second half of the 1980s provided empirical scientists with easier access to extensive data on order books. Biais et al. (1995) is an early study of the data flows on the newly (at that time) computerized Paris Bourse. Many subsequent papers offer complementary empirical findings and modelling perspectives, e.g., Gopikrishnan et al. (2000), Challet and Stinchcombe (2001), Maslov and Mills (2001), Bouchaud et al. (2002), Potters and Bouchaud (2003). In this chapter, we present a summary of some fundamental empirical facts. Basic statistical properties of limit order books, which can be observed from real data, are described and studied. Many variables crucial to a fine modelling of order flows and dynamics of order books are studied: Time of arrival of orders, placement of orders, size of orders, shape of order books, etc.

The markets we are dealing with are order-driven markets with no official market maker, in which orders are submitted in a double auction and executions follow price/time priority. In order to make the results we present both self-contained and reproducible, the statistics have been computed directly using our own database. The set of data that we have used in this chapter is detailed in Appendix B, which contains the precise description of all the data sets used throughout this book.

## 2.2 Time of Arrivals of Orders

We compute the empirical distribution for interarrival times – or durations – of market orders for the stock BNP Paribas using the data set described in Appendix B.2. The results are plotted in Figs 2.1 and 2.2, both in linear and log scale. It is clearly observed that the exponential fit is not a good one. We check however that the Weibull distribution fit is potentially a very good one. Weibull distributions have been suggested for example in Ivanov et al. (2004). Politi and Scalas (2008) also obtain good fits with $q$-exponential distributions.

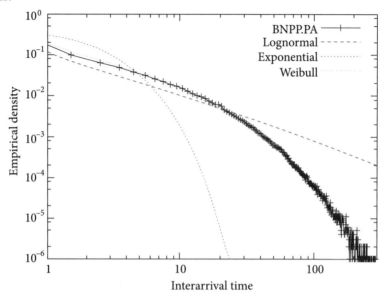

**Fig. 2.1** Distribution of interarrival times for stock BNPP.PA in log-scale. Extracted from Chakraborti et al. (2011a)

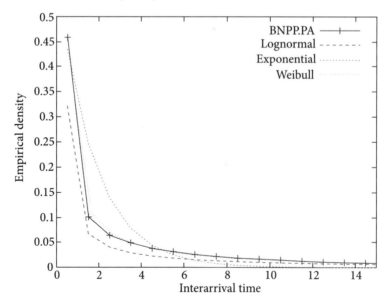

**Fig. 2.2** Distribution of interarrival times for stock BNPP.PA (Main body, linear scale). Extracted from Chakraborti et al. (2011a)

In the Econometrics literature, these observations of non-Poisson arrival times have given rise to a large trend of modelling of "irregular" financial data. Engle and Russell (1997) and Engle (2000) have introduced autoregressive condition duration or intensity models that may help modelling these processes of orders' submission (see Hautsch (2004) for a textbook treatment). Another trend of modelling that accounts for the non-exponential durations is based on the subordination of stochastic processes (Clark, 1973; Silva and Yakovenko, 2007; Huth and Abergel, 2012): A Poisson process with a change of time clock may be used to model financial data. Finally, this observation also leads to the use of *Hawkes processes* in financial modelling. These processes will be studied in Chapters 8 and 9.

We also compute using the same data the empirical distribution of the number of transactions in a given time period $\tau$. Results are plotted in Fig. 2.3.

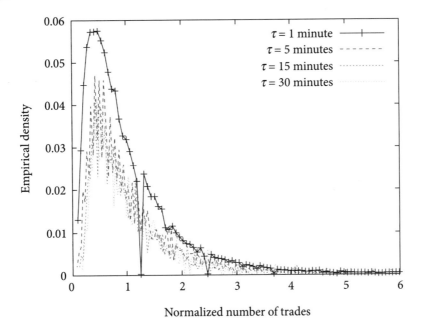

**Fig. 2.3** Distribution of the number of trades in a given time period $\tau$ for stock BNPP.PA. This empirical distribution is computed using data from 2007, October 1st until 2008, May 31st. Extracted from Chakraborti et al. (2011a)

It seems that the log-normal and the gamma distributions are both good candidates, however none of them really describes the empirical result, suggesting a complex structure of arrival of orders. A similar result on Russian stocks was presented in Dremin and Leonidov (2005).

Chapter 4 contains a more in-depth study of the arrival times of orders and their dependencies.

## 2.3 Volume of Orders

Empirical studies show that the distribution of order sizes is complex to characterize. A power-law distribution is often suggested. Gopikrishnan et al. (2000) and Maslov and Mills (2001) observe a power law decay with an exponent $1 + \mu \approx 2.3 - 2.7$ for market orders and $1 + \mu \approx 2.0$ for limit orders. Challet and Stinchcombe (2001) emphasize on a clustering property: Orders tend to have a "round" size in packages of shares, and clusters are observed around 100's and 1000's. As of today, no consensus emerges in proposed models, and it is plausible that such a distribution varies very wildly with products and markets.

In Fig. 2.4, we plot the distribution of volume of market orders for four different stocks. Quantities are normalized by their mean. Power-law coefficient is estimated by a Hill estimator [see e.g., Hill (1975); de Haan et al. (2000)]. We find a power law with exponent $1 + \mu \approx 2.7$ which confirms studies previously cited. Figure 2.5 displays the same distribution for limit orders (of all available limits).

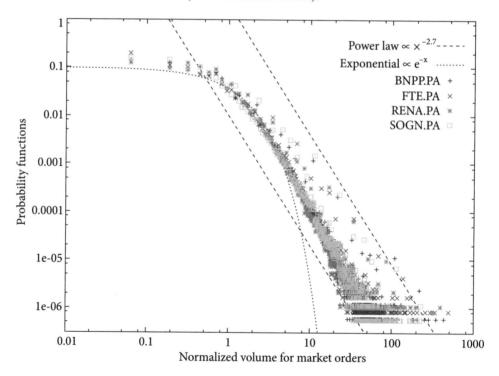

**Fig. 2.4** Distribution of volumes of market orders. Quantities are normalized by their mean. Extracted from Chakraborti et al. (2011a)

We find an average value of $1 + \mu \approx 2.1$, consistent with previous studies. However, we note that the power law is a poorer fit in the case of limit orders: Data normalized by their mean collapse badly on a single curve, and computed coefficients vary with stocks.

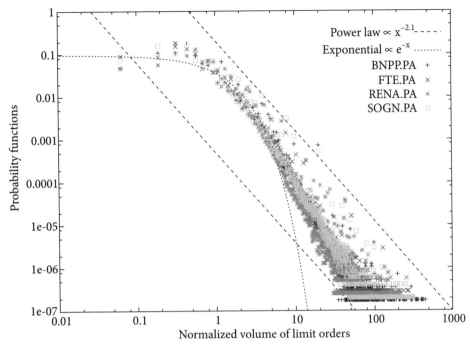

**Fig. 2.5** Distribution of normalized volumes of limit orders. Quantities are normalized by their mean. Extracted from Chakraborti et al. (2011a)

## 2.4 Placement of Orders

Bouchaud et al. (2002) observe a broad power-law placement around the best quotes on French stocks, confirmed in Potters and Bouchaud (2003) on US stocks. Observed exponents are quite stable across stocks, but exchange dependent: $1+\mu \approx 1.6$ on the Paris Bourse, $1+\mu \approx 2.0$ on the New York Stock Exchange, $1+\mu \approx 2.5$ on the London Stock Exchange. Mike and Farmer (2008) propose to fit the empirical distribution with a Student distribution with 1.3 degree of freedom.

We plot the distribution of the following quantity computed on our data set, i.e. using only the first five limits of the order book: $\Delta p = b_0(t-) - b(t)$ (resp. $a(t) - a_0(t-)$) if an bid (resp. ask) order arrives at price $b(t)$ (resp. $a(t)$), where $b_0(t-)$ (resp. $a_0(t-)$) is the best bid (resp. ask) before the arrival of this order. Results are plotted on Fig. 2.6 (in semilog scale) and 2.7 (in linear scale).

We observe that the empirical distribution of the placement of arriving limit orders is maximum at $\Delta p = 0$; i.e. at the same best quote. We also observe heavy tails in the distribution. Finally, we also observe an asymmetry in the empirical distribution: The left side is less broad than the right side. Since, the left side represent limit orders submitted *inside* the spread, this is expected: The left side of the distribution is linked to the spread distribution.

**14**     *Limit Order Books*

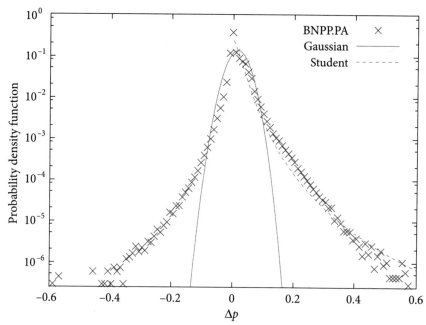

**Fig. 2.6**     Placement of limit orders using the same best quote reference in semilog scale. Data used for this computation is BNP Paribas order book from September 1st, 2007, until May 31st, 2008. Extracted from Chakraborti et al. (2011a)

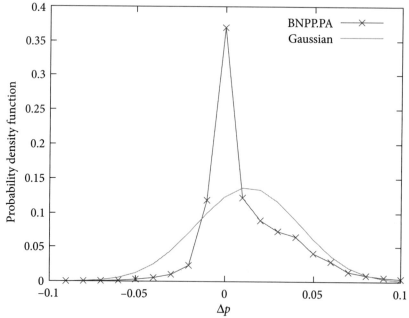

**Fig. 2.7**     Placement of limit orders using the same best quote reference in linear scale. Data used for this computation is BNP Paribas order book from September 1st, 2007, until May 31st, 2008. Extracted from Chakraborti et al. (2011a)

## 2.5 Cancellation of Orders

Challet and Stinchcombe (2001) show that the distribution of the average lifetime of limit orders fits a power law with exponent $1 + \mu \approx 2.1$ for cancelled limit orders, and $1 + \mu \approx 1.5$ for executed limit orders. Mike and Farmer (2008) find that in either case the exponential hypothesis (Poisson process) is not satisfied on the market.

We compute the average lifetime of cancelled and executed orders on our dataset. Since, our data does not include a unique identifier of a given order, we reconstruct life time orders as follows: Each time a cancellation is detected by the algorithm described in Appendix B.1, we go back through the history of limit order submission and look for a matching order with same price and same quantity. If an order is not matched, we discard the cancellation from our lifetime data. Results are presented in Figs 2.8 and 2.9. We observe a power law decay with coefficients $1 + \mu \approx 1.3 - 1.6$ for both cancelled and executed limit orders, with little variations among stocks. These results are a bit different from the ones presented in previous studies: Similar for executed limit orders, but our data exhibits a lower decay as for cancelled orders. Note that the observed cut-off in the distribution for lifetimes above 20000 seconds is due to the fact that we do not take into account execution or cancellation of orders submitted on a previous day.

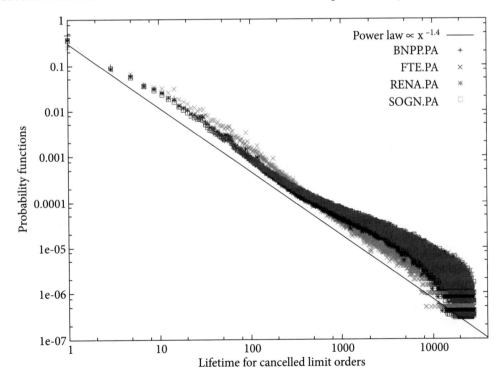

**Fig. 2.8** Distribution of estimated lifetime of cancelled limit orders. Extracted from Chakraborti et al. (2011a)

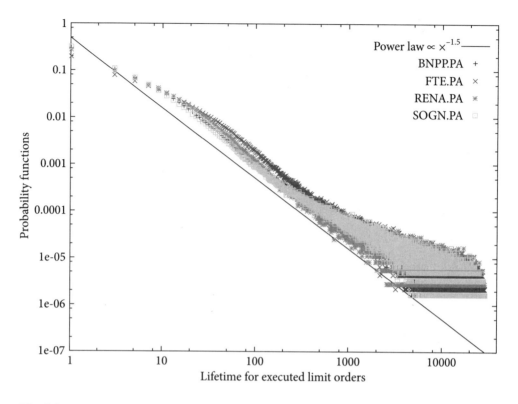

**Fig. 2.9** Distribution of estimated lifetime of executed limit orders. Extracted from Chakraborti et al. (2011a)

## 2.6 Average Shape of the Order Book

Contrary to what one might expect, the maximum of the average offered volume in an order book is located away from the best quotes (Bouchaud et al. 2002). Our data confirms this observation: The average quantity offered on the five best quotes grows with the level. This result is presented in Fig. 2.10. We also compute the average price of these levels in order to plot a cross-sectional graph similar to the ones presented in Biais et al. (1995). Our result is presented for stock BNP.PA in Fig. 2.11 and displays the expected shape. Results for other stocks are similar.

We find that the average gap between two levels is constant among the five best bids and asks (less than one tick for FTE.PA, 1.5 tick for BNPP.PA, 2.0 ticks for SOGN.PA, 2.5 ticks for RENA.PA). We also find that the average spread is roughly twice as large the average gap (factor 1.5 for FTE.PA, 2 for BNPP.PA, 2.2 for SOGN.PA, 2.4 for RENA.PA).

Chapter 3 presents more detailed results on the shape of the order book and its relation to the size of incoming orders.

Statistical Properties of Limit Order Books: A Survey    17

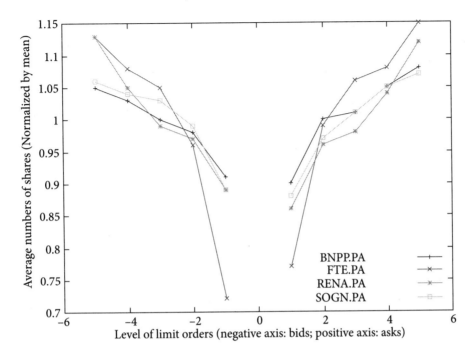

**Fig. 2.10** Average quantity offered in the limit order book. Extracted from Chakraborti et al. (2011a)

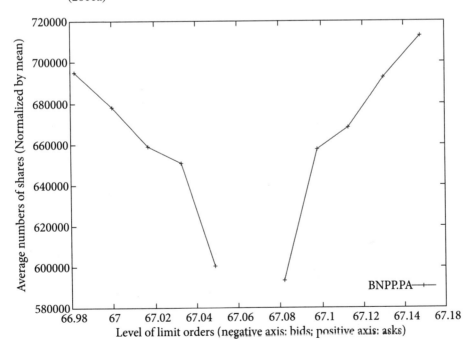

**Fig. 2.11** Average limit order book: price and depth. Extracted from Chakraborti et al. (2011a)

## 2.7 Intraday Seasonality

Activity on financial markets is of course not constant throughout the day. Figures 2.12 and 2.13 plot the (normalized) number of market and limit orders arriving in a 5-minute interval. It is clear that a U shape is observed (an ordinary least-square quadratic fit is plotted): the observed market activity is larger at the beginning and the end of the day, and more quiet around mid-day. Such a U-shaped curve is well-known, see Biais et al. (1995), for example. On our data, we observe that the number of orders on a 5-minute interval can vary with a factor 10 throughout the day.

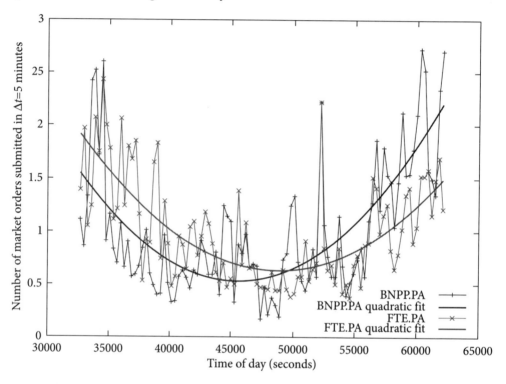

**Fig. 2.12** Normalized average number of market orders in a 5-minute interval. Extracted from Chakraborti et al. (2011a)

Challet and Stinchcombe (2001) note that the average number of orders submitted to the market in a period $\Delta T$ vary wildly during the day. The authors also observe that these quantities for market orders and limit orders are highly correlated. Such a type of intraday variation of the global market activity is a well-known fact, already observed in Biais et al. (1995), for example.

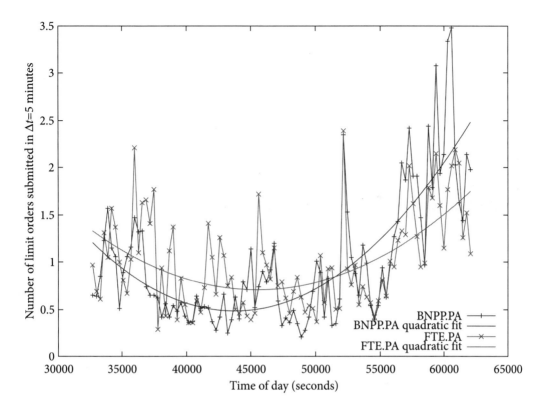

**Fig. 2.13** Normalized average number of limit orders in a 5-minute interval. Extracted from Chakraborti et al. (2011a)

## 2.8 Conclusion

In this introductory chapter, some elementary statistical features of limit order books have been produced, related to the size of order, the shape of the order book and the arrival times of orders. The next two chapters will delve further into two particular questions of interest: The connection between order sizes and the shape of the order book, and the mutual excitation of orders of different types.

# CHAPTER 3

# The Order Book Shape as a Function of the Order Size

## 3.1 Introduction

Elaborating on the results presented in Section 2.6, this chapter focuses on the shape of limit order books and the influence of the size of incoming orders. It confirms the theoretical findings of the models studied in Chapters 6 and 7.

## 3.2 Methodology

We use the order book data described in Appendix B.3. All movements on the first 10 limits of the ask side and the bid side of the order book are available, which allows us to reconstruct the evolution of the first limits of the order book during the day. Each trading day is divided into 12 thirty-minute intervals from 10am to 4pm. We obtain $T = 391$ intervals for each stock (three and a half days of trading are missing or incomplete in our dataset: January 15th, the morning of January 21st, February 18th and 19th). For each interval $t = 1, \ldots, T$, and for each stock $k = 1, \ldots, 14$, we compute the total number of limit orders $L^k(t)$ and market orders $M^k(t)$, and the average volume (in euros) of limit orders $V_L^k(t)$ and market orders $V_M^k(t)$. Table 3.1 gives the average number of orders and their volumes (the overline denotes the average over the time intervals: $\overline{L}^k = \frac{1}{T} \sum_{t=1}^{T} L^k(t)$, and similarly for other quantities). The lowest average activity is observed on UBIP.PA and LAGA.PA (which are the only two stocks in the sample with less than 200 market orders and 4000 limit orders in average). The highest activity is observed on BNPP.PA (which is the only stock with more than 600 market orders and 6000 limit orders on average). The smallest sizes of orders are observed on AIRP.PA (108.25 and 205.4 for market and limit orders), and the largest sizes are observed on AXA.PA (535.6 and 876.6 for market and limit orders).

**Table 3.1** Basic statistics on the number of orders and the average volumes of orders per 30-minute time interval for each stock

| Stock k | $\overline{M^k}$ (min, max) | $\overline{V^k_M}$ (min, max) | $\overline{L^k}$ (min, max) | $\overline{V^k_L}$ (min, max) |
|---|---|---|---|---|
| AIRP.PA | 290.9<br>59    1015 | 108.2<br>60.2    236.9 | 4545.4<br>972    24999 | 205.4<br>153.6    295.7 |
| ALSO.PA | 349.7<br>68    1343 | 181.3<br>75.7    310.8 | 5304.2<br>817    31861 | 289.2<br>234.5    389.6 |
| AXAF.PA | 521.8<br>119    2169 | 535.6<br>311.4    1037.4 | 4560.6<br>1162    20963 | 876.6<br>603.4    1648.3 |
| BNPP.PA | 773.7<br>160    4326 | 203.1<br>113.1    379.5 | 6586.0<br>946    42939 | 277.6<br>189.2    518.7 |
| BOUY.PA | 218.3<br>38    1021 | 227.4<br>113.1    421.9 | 4544.5<br>652    25546 | 369.7<br>274.5    475.2 |
| CARR.PA | 309.5<br>49    1200 | 275.3<br>156.1    517.3 | 4391.4<br>813    14752 | 513.3<br>380.8    879.6 |
| DANO.PA | 385.7<br>86    2393 | 214.4<br>112.6    820.3 | 4922.1<br>1372    18186 | 393.1<br>286.1    537.4 |
| LAGA.PA | 140.8<br>22    544 | 201.5<br>87.2    376.8 | 3429.0<br>632    11319 | 338.2<br>219.0    504.0 |
| MICP.PA | 301.2<br>45    1114 | 137.9<br>74.4    235.4 | 5033.5<br>1012    23799 | 240.7<br>178.0    315.1 |
| PEUP.PA | 294.9<br>57    1170 | 333.2<br>159.1    657.7 | 3790.6<br>967    14053 | 536.7<br>316.8    828.3 |
| RENA.PA | 463.2<br>103    1500 | 266.9<br>133.7    477.2 | 4957.6<br>1383    27851 | 384.3<br>283.1    539.8 |
| SASY.PA | 494.0<br>106    1722 | 241.4<br>128.7    511.7 | 4749.1<br>986    22373 | 421.2<br>311.1    670.2 |
| SGEF.PA | 366.1<br>70    1373 | 188.4<br>107.7    452.3 | 5309.9<br>983    21372 | 353.4<br>274.2    515.1 |
| UBIP.PA | 153.3<br>18    998 | 384.9<br>198.4    675.1 | 1288.9<br>240    7078 | 754.0<br>451.4    1121.0 |

We also compute the average cumulative order book shape at 1 to 10 ticks from the best opposite side. Denoting by $P^B(t)$ and $P^A(t)$ the best bid and ask prices at time $t$, the *average cumulative depth* $B^k_i(t)$ is the cumulative quantity available in the order book for stock $k$ in the price range $\{P^B(t)+1, \ldots, P^B(t)+i\}$ (in ticks) for ask limit orders, or in the price range $\{P^A(t)-i, \ldots, P^A(t)-1\}$ for bid limit orders, averaged over time during interval $t$. For $i \leq 10$, our data is always complete since, the first ten limits are available. For larger $i$ however, we may not have the full data: $B^k_{10+j}(t)$ is not exact if the spread reaches a level lower or equal to $j$ ticks during interval $t$. Hence, we impose that $i \leq 10$ in the following empirical analysis. Figure 3.1 plots the empirical average shapes $\overline{B}^k_i = \frac{1}{T}\sum_{t=1}^{T} B^k_i(t)$ (arbitrarily scaled to $\overline{B}^k_{10} = 1$ for easier comparison).

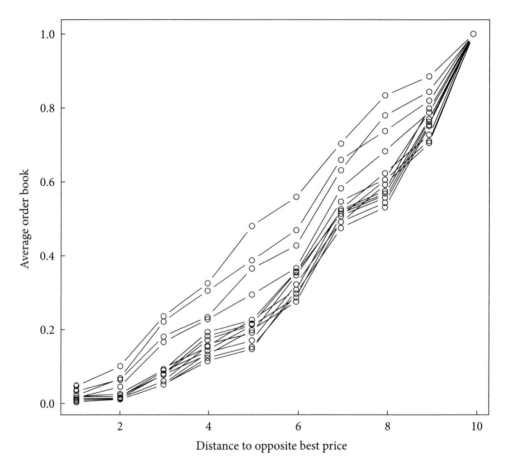

**Fig. 3.1** Mean-scaled shapes of the cumulative order book as a function of the distance (in ticks) from the opposite best price for the 14 stocks studied. Extracted from Muni Toke (2015)

## 3.3 The Regression Model

We investigate the influence of the number of limit orders $L^k(t)$ and their average size $V^k_L(t)$ on the depth on the order book at the first limits with a regression model. We wish to study this relationship with the number of limit orders and their size all others things being equal, i.e. the global market activity being held constant. A fairly natural proxy for the market activity is the traded volume (transactions) per period. The total volume of market orders submitted during interval $t$ for the stock $k$ is $M^k(t)V_M^k(t)$. We may also consider the total volume of incoming limit orders $L^k(t)V_L^k(t)$.

Therefore, we have the following regression model for some measure of the book depth $B^k(t)$:

$$B^k(t) = \beta_1 L^k(t) + \beta_2 V_L^k(t) + \beta_3 L^k(t) V_L^k(t) + \beta_4 M^k(t) V_M^k(t) + \epsilon_{k,t}. \tag{3.1}$$

We test this regression using $B^k(t) = B_5^k(t)$ (cumulative depth up to 5 ticks away from the best opposite price), then, using $B^k(t) = B_{10}^k(t)$ (cumulative depth up to 10 ticks away from the best opposite price) and finally using $B^k(t) = B_{10}^k(t) - B_5^k(t)$ (cumulative volume between 6 and 10 ticks away from the best opposite price). For each of these three models, we provide results with or without the interaction term $L^k V_L^k$. All models are estimated as panel regression models with fixed effects, i.e. the error term $\epsilon_{k,t}$ is the sum of a non-random stock specific component $\delta_k$ (fixed effect) and a random component $\eta_{k,t}$. While the regression coefficients are stock-independent, the variables $\delta_k$ translate the idiosyncratic characteristics of each stock.

Note that we use data from 10am to 4pm each day, in order to avoid very active periods, right after the opening of the market or before its closing. By concentrating on the heart of the trading day, we focus on smoother variations of the studied variables. However, even with this restriction, the data nonetheless exhibits some intraday seasonality, such as the well-known U-shaped curve of the number of submissions of orders (see Chapter 2), or the fact that the order book seems to grow slightly fuller during the day and decline in the end. This intraday seasonality may be observed by computing, for each stock, the (intraday) seasonal means of the variables of the model, e.g., the average of a given variable for a given stock at a given time of the day across the whole sample. This is illustrated for example in Fig. 3.2 and Fig. 3.3 where we plot, for the 14 stocks studied, the (mean-scaled) seasonal averages of the number of submitted market orders and the (mean-scaled) seasonal averages of the cumulative size of the order book up to the tenth limit.

For the sake of completeness, we run the statistical regressions defined at Eq. (3.1) both on raw data and on deseasonalized data (by subtracting the seasonal mean). Results are given in Table 3.2 in the first case, and in Table 3.3 in the latter one.

In all cases, irrespective of the presence of the interaction term $L^k V_L^k$ and irrespective of the deseasonalization, we observe a positive relationship between the order book depth and the average size of limit orders $V_L^k$, and a negative relationship between the order book depth and the number of limit orders $L^k$. Note also that the estimated $\beta$'s for these two quantities are all significant to the highest level in all the cases using deseasonalized data, and to the 0.2% for all but one case using the raw data (2% level in this worst case).

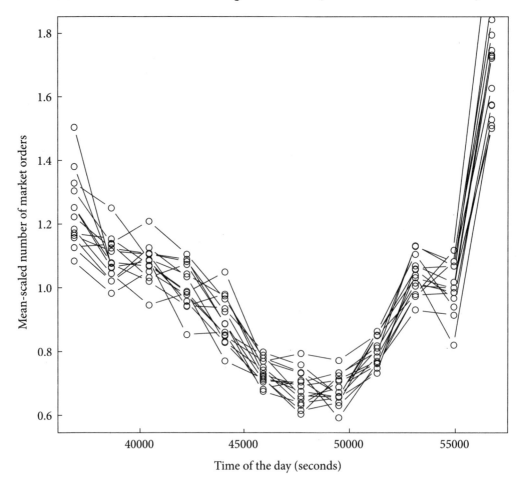

**Fig. 3.2** Mean-scaled number of market orders as a function of the time of day (in seconds) for the 14 stocks studied. Extracted from Muni Toke (2015)

Therefore, all others things being equal, it thus appears we have identified the following effect: For a given total volume of arriving limit and market orders $L^k V_L^k$ and $M^k V_M^k$, the relative size of the limit orders has a strong influence on the average shape of the order book: The larger the arriving orders are, the deeper the order book is; the more there are arriving limit orders, the shallower the order book is. The average shape of the order book is deeper when a few large limit orders are submitted, than when many small limit orders are submitted. We will provide in Chapter 7, and in particular in Section 7.4, a theoretical order book model that precisely describes this effect.

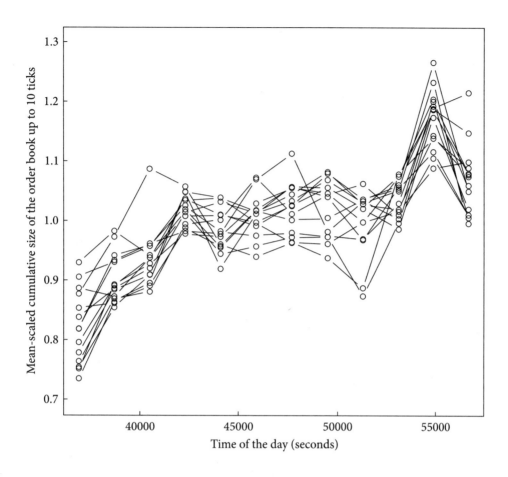

**Fig. 3.3** Mean-scaled shapes of the cumulative order book up to 10 ticks away from the opposite best price as a function of the time of day (in seconds), for the 14 stocks studied. Extracted from Muni Toke (2015)

As a side comment, we may also look at the influence of the global volume of trades $M^k V_M^k$. Using raw data, we observe a positive influence of this term on the depth of the order book. At a first glance, this is in line with the phenomenon observed for example in Ns and Skjeltorp (2006), where a positive relationship between the number of trades and the order book slope of the first 5 limits of the order book is exhibited. This apparent similarity is to be taken with great caution, since, the slope and the depth are two different variables: their relationship and the possible link between their evolutions is not known. Note in particular that the positive influence of the global volume of trades does not hold using deseasonalized data in our sample. A second observation made by Ns and Skjeltorp (2006) is that the influence of market activity on the order book depth is much stronger closer to the spread, and then, decreases when taking into account further limits. We also observe this phenomenon using raw data: The lower panel of Table 3.2 shows that this influence is

**Table 3.2** Panel regression results for the models defined in Eq. (3.1), using raw data, for $B_{k,t}^5$ (top panel), $B_{k,t}^{10}$ (middle panel) and $B_{k,t}^{10} - B_{k,t}^5$ (lower panel). In all case, data has 14 stocks and $T = 391$ time intervals, i.e. 5474 points

| $B_{k,t}^5$ | Without $L^k(t)V_M^k(t)$ | | | | | With $L^k(t)V_M^k(t)$ | | | | |
|---|---|---|---|---|---|---|---|---|---|---|
| Variables | Est. | Std. Err. | t-value | p-value | Sig. | Est. | Std. Err. | t-value | p-value | Sig. |
| $L^k(t)$ | -0.24455434 | 0.02008400 | -12.177 | < 2.2e-16 | *** | -8.65511e-02 | 3.60046e-02 | -2.4000 | 0.01643 | * |
| $V_M^k(t)$ | 8.70817318 | 0.75084050 | 11.598 | < 2.2e-16 | *** | 1.0374e+01 | 8.1288e-01 | 12.7622 | < 2.2e-16 | *** |
| $L^k(t)V_M^k(t)$ | — | — | — | — | — | -4.8483e-04 | 9.1924e-05 | -5.2743 | 1.384e-07 | *** |
| $L^k(t)V_M^k(t)$ | 0.00082725 | 0.00079296 | 11.132 | < 2.2e-16 | *** | 1.0794e-02 | 8.7448e-04 | 12.3430 | < 2.2e-16 | *** |
| $R^2$ | 0.079275 | | | | | 0.083945 | | | | |
| F-statistic | 156.617 | | | < 2.2e-16 | *** | 124.994 | | | < 2.2e-16 | *** |

| $B_{k,t}^{10}$ | Without $L^k(t)V_M^k(t)$ | | | | | With $L^k(t)V_M^k(t)$ | | | | |
|---|---|---|---|---|---|---|---|---|---|---|
| Variables | Est. | Std. Err. | t-value | p-value | Sig. | Est. | Std. Err. | t-value | p-value | Sig. |
| $L^k(t)$ | -0.4576647 | 0.0439167 | -10.4212 | < 2.2e-16 | *** | -0.25599592 | 0.07895182 | -3.2424 | 0.001192 | ** |
| $V_M^k(t)$ | 39.4968315 | 1.6418268 | 24.0566 | < 2.2e-16 | *** | 41.62260408 | 1.78046702 | 23.3774 | < 2.2e-16 | *** |
| $L^k(t)V_M^k(t)$ | — | — | — | — | — | -0.00061866 | 0.00020134 | -3.0727 | 0.002132 | ** |
| $L^k(t)V_M^k(t)$ | 0.0097127 | 0.0017339 | 5.6016 | 2.228e-08 | *** | 0.01222198 | 0.00191540 | 6.3809 | 1.906e-10 | *** |
| $R^2$ | 0.14202 | | | | | 0.14303 | | | | |
| F-statistic | 301.1 | | | < 2.2e-16 | *** | 228.534 | | | < 2.2e-16 | *** |

| $B_{k,t}^{10} - B_{k,t}^5$ | Without $L^k(t)V_M^k(t)$ | | | | | With $L^k(t)V_M^k(t)$ | | | | |
|---|---|---|---|---|---|---|---|---|---|---|
| Variables | Est. | Std. Err. | t-value | p-value | Sig. | Est. | Std. Err. | t-value | p-value | Sig. |
| $L^k(t)$ | -0.21311036 | 0.02995514 | -7.1143 | 1.27e-12 | *** | -0.16948494 | 0.05389411 | -3.1448 | 0.001671 | ** |
| $V_M^k(t)$ | 30.78865834 | 1.11987302 | 27.4930 | < 2.2e-16 | *** | 31.24851002 | 1.21538277 | 25.7108 | < 2.2e-16 | *** |
| $L^k(t)V_M^k(t)$ | — | — | — | — | — | -0.00013383 | 0.00013744 | -0.9737 | 0.330234 | |
| $L^k(t)V_M^k(t)$ | 0.00088543 | 0.00118269 | 0.7487 | 0.4541 | | 0.00142825 | 0.00130749 | 1.0924 | 0.274723 | |
| $R^2$ | 0.1491 | | | | | 0.14925 | | | | |
| F-statistic | 318.743 | | | < 2.2e-16 | *** | 239.292 | | | < 2.2e-16 | *** |

**Table 3.3** Panel regression results for the models defined in Eq. (3.1), using deseasonalized data, for $B_{k,t} = B_{k,t}^5$ (top panel), $B_{k,t} = B_{k,t}^{10}$ (middle panel) and $B_{k,t} = B_{k,t}^{10} - B_{k,t}^5$ (lower panel). In all case, data has 14 stocks and $T = 391$ time intervals, i.e. 5474 points

| $B_{k,t}^5$ | Without $L^k(t)V_M^k(t)$ | | | | | With $L^k(t)V_M^k(t)$ | | | | |
|---|---|---|---|---|---|---|---|---|---|---|
| Variables | Est. | Std Err. | t-value | p-value | Sig. | Est. | Std Err. | t-value | p-value | Sig. |
| $L^k(t)$ | -0.2693183 | 0.0199388 | -13.5072 | < 2.2e-16 | *** | -0.2712167 | 0.0198836 | -13.6402 | < 2.2e-16 | *** |
| $V_M^k(t)$ | 14.8579856 | 0.6885427 | 21.5789 | < 2.2e-16 | *** | 15.2819415 | 0.6905058 | 22.1315 | < 2.2e-16 | *** |
| $L^k(t)V_M^k(t)$ | — | — | — | — | — | -0.0017696 | 0.0003085 | -5.7361 | 1.021e-08 | *** |
| $L^k(t)V_M^k(t)$ | -0.0167731 | 0.0036252 | -4.6268 | 3.799e-06 | *** | -0.0064651 | 0.0040367 | -1.6016 | 0.1093 | |
| $R^2$ | 0.10337 | | | | | 0.10875 | | | | |
| F-statistic | 209.718 | | | < 2.2e-16 | *** | 166.433 | | | < 2.2e-16 | *** |

| $B_{k,t}^{10}$ | Without $L^k(t)V_M^k(t)$ | | | | | With $L^k(t)V_M^k(t)$ | | | | |
|---|---|---|---|---|---|---|---|---|---|---|
| Variables | Est. | Std Err. | t-value | p-value | Sig. | Est. | Std Err. | t-value | p-value | Sig. |
| $L^k(t)$ | -0.6663429 | 0.0423425 | -15.7370 | < 2.2e-16 | *** | -0.67127949 | 0.04216152 | -15.9216 | < 2.2e-16 | *** |
| $V_M^k(t)$ | 50.9522688 | 1.4622037 | 34.8462 | < 2.2e-16 | *** | 52.05475178 | 1.46416246 | 35.5526 | < 2.2e-16 | *** |
| $L^k(t)V_M^k(t)$ | — | — | — | — | — | -0.00460179 | 0.00065416 | -7.0347 | 2.243e-12 | *** |
| $L^k(t)V_M^k(t)$ | -0.0429541 | 0.0076985 | -5.5795 | 2.528e-08 | *** | -0.01614842 | 0.00855951 | -1.8866 | 0.05927 | . |
| $R^2$ | 0.20566 | | | | | 0.2128 | | | | |
| F-statistic | 470.96 | | | < 2.2e-16 | *** | 368.73 | | | < 2.2e-16 | *** |

| $B_{k,t}^{10} - B_{k,t}^5$ | Without $L^k(t)V_M^k(t)$ | | | | | With $L^k(t)V_M^k(t)$ | | | | |
|---|---|---|---|---|---|---|---|---|---|---|
| Variables | Est. | Std Err. | t-value | p-value | Sig. | Est. | Std Err. | t-value | p-value | Sig. |
| $L^k(t)$ | -0.3970246 | 0.0293712 | -13.5175 | < 2.2e-16 | *** | -0.4000628 | 0.0292738 | -13.6662 | < 2.2e-16 | *** |
| $V_M^k(t)$ | 36.0942832 | 1.0142678 | 35.5865 | < 2.2e-16 | *** | 36.7728102 | 1.0166060 | 36.1721 | < 2.2e-16 | *** |
| $L^k(t)V_M^k(t)$ | — | — | — | — | — | -0.0028322 | 0.0004542 | -6.2356 | 4.84e-10 | *** |
| $L^k(t)V_M^k(t)$ | -0.0261809 | 0.0053401 | -4.9027 | 9.728e-07 | *** | -0.0096833 | 0.0059431 | -1.6293 | 0.1033 | |
| $R^2$ | 0.20463 | | | | | 0.21026 | | | | |
| F-statistic | 467.995 | | | < 2.2e-16 | *** | 363.153 | | | < 2.2e-16 | *** |

not significant anymore if we take only the "furthest" limits, i.e. the limits between 6 and 10 ticks away from the best price. This is however unclear with deseasonalized data.

## 3.4 Conclusion

The statistics presented in this chapter provide clear evidence of the influence on the shape of the limit order book of the size and numbers of limit orders: Larger limit orders and fewer limit orders lead to fatter order books. Such empirical findings will support the theoretical work presented in Chapter 7.

CHAPTER 4

# Empirical Evidence of Market Making and Taking

## 4.1 Introduction

In this chapter, we present empirical studies that shed some new light on the dependency structure of order arrival times, in particular, on the mutual and self-excitations of limit and market orders and, to a lesser extent, of cancellation orders. Examples for various assets and markets (mostly equities, but also bond and index futures) are provided. These empirical studies lay the ground for the advanced mathematical models studied in Chapters 8 and 9.

## 4.2 Re-introducing Physical Time

As seen in Chapter 2, the Poisson hypothesis for the arrival times of orders of different kinds does not stand under careful scrutiny. However, the study of arrival times of orders in an order book has not been a primary focus in the first attempts at order book modelling. Toy models leave this dimension aside when trying to understand the complex dynamics of an order book. In many order-driven market models (Cont and Bouchaud, 2000; Lux and Marchesi, 2000; Alfi et al. 2009a), and in some order book models as well (Preis et al., 2006), a time step in the model is an arbitrary unit of time during which many events may happen. We may call that clock *aggregated time*. In most order book models (Challet and Stinchcombe, 2001; Mike and Farmer, 2008), one order is simulated per time step with given probabilities, i.e. these models use the clock known as *event time*. In the simple case where these probabilities are time-homogeneous and independent of the state of the model, such a time treatment is equivalent to the assumption that order flows are homogeneous Poisson processes. A likely reason for the use of event time in order book modelling – leaving aside the fact that models can be sufficiently complicated without adding another dimension – is that many puzzling empirical observations can be made in event time (e.g. autocorrelation of the signs of limit and market orders) or in aggregated time (e.g. volatility clustering) (see Chapter 2).

However, it is clear that *physical* time has to be taken into account for the modelling of a realistic order book model. For example, market activity varies widely, and intraday seasonality is often observed as a well known U-shaped pattern. Even for a short time scale model – a few minutes, a few hours – durations of orders (i.e. time intervals between orders) are very broadly distributed. Hence, the Poisson assumption and its exponential distribution of arrival times have to be discarded, and models must take into account the way these irregular flows of orders affect the empirical properties studied on order books.

Let us give one illustration. On Fig. 4.1, we have plotted examples of the empirical distribution function of the observed spread in event time (i.e. spread is measured each time an event happens in the order book), and in physical (calendar) time (i.e. measures are weighted by the time interval during which the order book is idle).

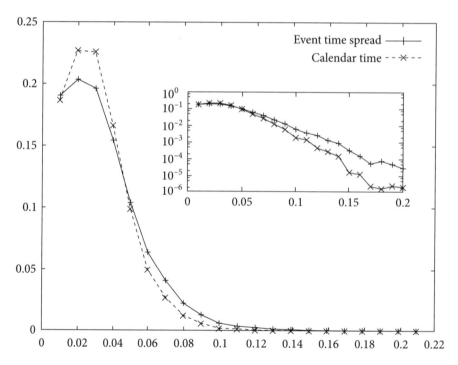

**Fig. 4.1** Empirical distribution function of the bid-ask spread in event time and in physical (calendar) time. In inset, same data using a semi-log scale. This graph has been computed with 15 four-hour samples of tick data on the BNPP.PA stock. Extracted from Muni Toke (2011)

It appears that the frequencies of the most probable values of the time-weighted distribution are higher than in the event time case. Symmetrically, the frequencies of the least probable events are even smaller when physical time is taken into account. This tells us a few things about the dynamics of the order book, which could be summarized as follows: The wider the spread, the faster its tightening. We can get another insight of this empirical property by measuring on our data the average waiting time before the next event, conditionally on the spread size. When computed on the lower one-third-quantile (small spread), the average waiting time is 320 milliseconds. When computed on the upper one-third-quantile (large spread), this average waiting time is 200 milliseconds. These observations complement some of the ones that can be found in Biais et al. (1995).

## 4.3 Dependency Properties of Inter-arrival Times

### 4.3.1 Empirical evidence of market making

A first idea for an enhanced model of order flows is based on the following observation: Once a market order has been placed, the next limit order is likely to take place faster than usual. To illustrate this, we compute for several assets:

- the empirical probability distribution of the time intervals of the counting process of all orders (limit orders and market orders mixed), i.e. the time step between any order book event (other than cancellation)
- and the empirical probability distribution of the time intervals between a market order and the immediately following limit order.

If an independent Poisson assumption held, then these empirical distributions should be identical. However, we observe a very high peak for short time intervals in the second case. The first moment of these empirical distributions is significant: For the studied assets, we find that the average time between a market order and the following limit order is 1.3 (BNPP.PA) to 2.6 (LAGA.PA) times shorter than the average time between two random consecutive events.

On the graphs shown in Fig. 4.2, we plot the full empirical distributions for four of the five studied assets[1]. We observe their broad distribution and the sharp peak for the shorter times: on the Footsie future market for example, 40% of the measured time steps between consecutive events are less that 50 milliseconds; this figure jumps to nearly 70% when considering only market orders and their following limit orders. This observation is an evidence for some sort of market making behaviour of some participants on those markets. It seems that the submission of market orders is monitored and triggers automatic limit orders that add volumes in the order book (and not far from the best quotes, since, we only monitor the five best limits).

In order to confirm this finding, we perform non-parametric statistical tests on the measured data. For all studied markets, omnibus Kolmogorov–Smirnov and Cramer–von

---
[1] Observations are identical on all the studied assets.

Mises tests performed on random samples establish that the considered distributions are statistically different. If assuming a common shape, a Wilcoxon-Mann-Withney U test clearly states that the distribution of time intervals between a market order and the following limit order is clearly shifted to the left compared to the distributions of time intervals between any orders, i.e. the average "limit following market" reaction time is shorter than the average time interval between random consecutive orders.

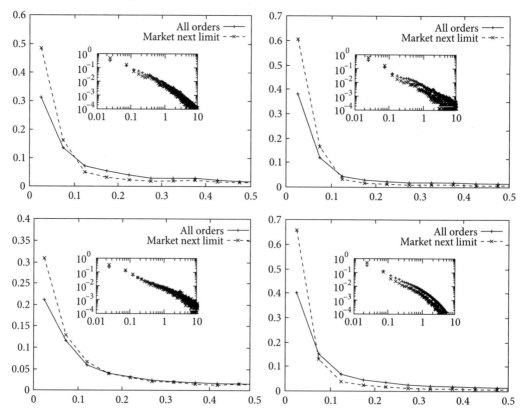

**Fig. 4.2** Empirical distributions of the time intervals between two consecutive orders (any type, market or limit) and of the time intervals between a market order and the immediately following limit order. X-axis is scaled in seconds. In insets, same data using a log-log scale. Studied assets: BNPP.PA (top left), LAGA.PA (top right), FEIZ9 (bottom left), FFIZ9 (bottom right). Extracted from Muni Toke (2011)

Note that there is no systematic link between the sign of the market order and the sign of the following limit order. For example for the BNP Paribas (resp. Peugeot and Lagardere) stock, they have the same sign in 48.8% (resp. 51.9% and 50.7%) of the observations. And more interestingly, the "limit following market" property holds regardless of the side on which the following limit order is submitted. On Fig. 4.3, we have plotted the empirical distributions of time intervals between a market order and the following limit order, conditionally on the side of the limit order: The same side as the market order or the opposite one. It appears for all studied assets that both distributions are roughly identical.

In other words, we cannot clearly distinguish on the data if liquidity is added where the market order has been submitted or on the opposite side. Therefore, we do not infer any empirical property of placement: When a market order is submitted, the intensity of the limit order process increases *on both sides* of the order book.

This effect we have thus identified is a phenomenon of liquidity replenishment of an order book after the execution of a trade. The fact that it is a bilateral effect makes its consequences similar to that of a market making strategy, even though there is no official market maker involved on every studied market. This market making behaviour will be characterized in a bit more detailed fashion in Section 4.4.

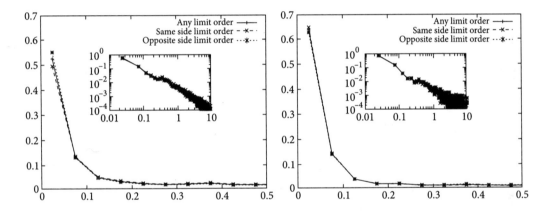

**Fig. 4.3** Empirical distributions of the time intervals between a market order and the immediately following limit order, whether orders have been submitted on the same side and on opposite sides. X-axis is scaled in seconds. In insets, same data using a log-log scale. Studied assets: BNPP.PA (left), LAGA.PA (right). Extracted from Muni Toke (2011)

### 4.3.2 A reciprocal effect?

We now check if a similar or opposite distortion is to be found on market orders when they follow limit orders. To investigate this, we compute the "reciprocal" measures for all our studied assets:

- the empirical distribution of the time intervals of the counting process of all orders (limit orders and market orders mixed), i.e. the time step between any order book event (other than cancellation)
- and the empirical distribution of the time step between a market order and the previous limit order.

If an independent assumption held, then these empirical distribution should be identical. Results for four assets are shown on Fig. 4.4.

Contrary to the previous case, no effect is very easily interpreted. For the three stocks [BNPP.PA, LAGA.PA and PEUP.PA (not shown)], it seems that the empirical distribution is less peaked for small time intervals, but the difference is much less important than in the previous case. As for the FEI and FFI markets, the two distributions are even more closer.

Non-parametric tests confirm these observations. Performed on data from the three equity markets, Kolmogorov tests indicate different distributions and Wilcoxon tests enforce the observation that time intervals between a limit order and a following market order are stochastically larger than time intervals between unidentified orders.

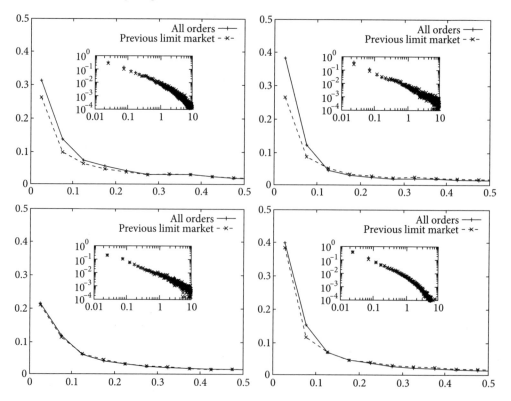

**Fig. 4.4** Empirical distributions of the time intervals between two consecutive orders (any type, market or limit) and of the time intervals between a limit order and an immediately following market order. In insets, same data using a log-log scale. Studied assets: BNPP.PA (top left), LAGA.PA (top right), FEIZ9 (bottom left), FFIZ9 (bottom right). Extracted from Muni Toke (2011)

As for the future markets on Footsie (FFI) and 3-month Euribor (FEI), Kolmogorov tests do not indicate differences in the two observed distributions, and the result is confirmed by a Wilcoxon test that concludes at the equality of the means. However, these results have been obtained by considering *all* limit orders and *all* market orders, whereas a market taking behaviour is more likely to be observed when considering limit orders that instantaneously change the price. In the next section, we make such a distinction, and provide convincing empirical evidences of this effect.

## 4.4 Further Insight into the Dependency Structure

As advertised, we now consider the events with more caution, explicitly distinguishing between *aggressive* orders that imediately change the best bid or ask price, and *passive* orders that do not.

Table 4.1 summarizes the definitions and notations for the different types of events used in this section.

**Table 4.1** Event types definitions

| Notation | Definition |
|---|---|
| $M, L, C, O$ | market order, limit order, cancellation, any order. |
| $M_{buy}, M_{sell}$ | buy/sell market order. |
| $M^0_{buy}, M^0_{sell}$ | buy/sell market order that does not change the mid price: i.e. order quantity < best ask/bid available quantity. |
| $M^1_{buy}, M^1_{sell}$ | buy/sell market order that changes the mid price: i.e. order quantity ≥ best ask/bid available quantity. |
| $L_{buy}, L_{sell}$ | buy/sell limit order. |
| $L^0_{buy}, L^0_{sell}$ | buy/sell limit order that does not change the mid price: i.e. order price ≤ / ≥ best bid/ask price. |
| $L^1_{buy}, L^1_{sell}$ | buy/sell limit order that changes the mid price: i.e. order price > / < best bid/ask price. |
| $C_{buy}, C_{sell}$ | buy/sell cancellation. |
| $C^0_{buy}, C^0_{sell}$ | buy/sell cancellation that does not change the mid price: i.e. partial cancellation at best bid/ask limit or cancellation at another limit. |
| $C^1_{buy}, C^1_{sell}$ | buy/sell cancellation that changes the mid price: i.e. total cancellation of best bid/ask limit order. |
| $M^0, L^0, C^0, O^0$ | market order, limit order, cancellation, any order, that does not change the mid price. |
| $M^1, L^1, C^1, O^1$ | market order, limit order, cancellation, any order, that changes the mid price. |

Table 4.2 summarizes the average daily numbers of each type of events.

**Table 4.2** Event occurrences statistics summary

|         | $L_{buy}$ | $L_{sell}$ | $L$   | $C_{buy}$ | $C_{sell}$ | $C$   | $M_{buy}$ | $M_{sell}$ | $M$   | $O$    |
|---------|-----------|------------|-------|-----------|------------|-------|-----------|------------|-------|--------|
| Average | 24020     | 24219      | 48239 | 20328     | 20591      | 40919 | 3870      | 3876       | 7764  | 96904  |
| Min     | 8804      | 8883       | 17687 | 7062      | 7410       | 14472 | 1575      | 1481       | 3056  | 36433  |
| Max     | 44321     | 46123      | 90444 | 41296     | 41075      | 82371 | 7665      | 7321       | 14986 | 187801 |

The statistics clearly show some symmetry between the buy and sell sides. The numbers of limit orders and cancellations are comparable, both are significantly higher than the number of market orders.

Going further, one can empirically estimate the conditional probabilities of occurrence for events of all types. Table 4.3 shows such probabilities of occurrence of an event of type $j$ (in column) conditional to the fact that the last observed event is of type $i$ (in row). The last row represents the unconditional probabilities of each type of events.

**Table 4.3** Conditional probabilities (in %) of occurrences per event type

|              | $L^0_{buy}$ | $L^0_{sell}$ | $C^0_{buy}$ | $C^0_{sell}$ | $M^0_{buy}$ | $M^0_{sell}$ | $L^1_{buy}$ | $L^1_{sell}$ | $C^1_{buy}$ | $C^1_{sell}$ | $M^1_{buy}$ | $M^1_{sell}$ |
|--------------|-------------|--------------|-------------|--------------|-------------|--------------|-------------|--------------|-------------|--------------|-------------|--------------|
| $|L^0_{buy}$  | 41.37 | 9.64  | 16.00 | 22.40 | 2.90  | 1.58  | 2.35  | 1.12  | 0.02  | 1.08  | 1.39  | 0.16  |
| $|L^0_{sell}$ | 9.61  | 41.79 | 21.95 | 16.12 | 1.61  | 2.96  | 1.02  | 2.29  | 1.05  | 0.02  | 0.15  | 1.44  |
| $|C^0_{buy}$  | 17.91 | 25.88 | 40.67 | 5.98  | 1.39  | 1.74  | 1.20  | 2.34  | 1.49  | 0.37  | 0.56  | 0.47  |
| $|C^0_{sell}$ | 25.18 | 17.98 | 6.04  | 41.30 | 1.79  | 1.42  | 2.08  | 1.27  | 0.37  | 1.49  | 0.51  | 0.60  |
| $|M^0_{buy}$  | 22.17 | 5.33  | 4.75  | 9.94  | 34.64 | 0.70  | 7.68  | 0.65  | 0.55  | 1.31  | 11.86 | 0.42  |
| $|M^0_{sell}$ | 5.60  | 21.14 | 10.61 | 5.01  | 0.72  | 34.32 | 0.53  | 7.19  | 1.48  | 1.10  | 0.42  | 11.88 |
| $|L^1_{buy}$  | 32.39 | 8.06  | 0.21  | 25.27 | 4.84  | 5.58  | 1.42  | 1.57  | 5.80  | 1.77  | 2.44  | 10.65 |
| $|L^1_{sell}$ | 7.65  | 29.94 | 26.04 | 0.22  | 5.63  | 5.62  | 1.39  | 1.36  | 1.42  | 5.39  | 12.37 | 2.96  |
| $|C^1_{buy}$  | 25.02 | 19.09 | 35.70 | 4.96  | 0.96  | 0.67  | 8.34  | 3.59  | 0.72  | 0.35  | 0.48  | 0.12  |
| $|C^1_{sell}$ | 21.48 | 23.28 | 5.42  | 34.70 | 0.76  | 1.16  | 3.20  | 7.88  | 0.63  | 0.75  | 0.18  | 0.57  |
| $|M^1_{buy}$  | 28.27 | 9.60  | 7.38  | 28.12 | 3.11  | 1.02  | 11.52 | 7.98  | 0.90  | 0.87  | 0.67  | 0.55  |
| $|M^1_{sell}$ | 11.83 | 23.05 | 33.36 | 7.24  | 1.04  | 3.13  | 6.79  | 9.34  | 1.05  | 1.81  | 0.66  | 0.70  |
| $|O$          | 22.82 | 22.93 | 19.80 | 20.03 | 2.99  | 3.00  | 2.07  | 2.12  | 0.85  | 0.88  | 1.27  | 1.26  |

Results in Table 4.3 are rather symmetric: No significant difference is observed between the buy and sell side. Therefore, their interpretation is detailed only in the case of buy orders:

$L^0_{buy}$: Reinforces the consensus that the stock is not moving down. This increases the probability of $L^0_{buy}$.

$C_{buy}^0$: Decreases the available liquidity on the buy side. Other participants may feel less comfortable posting buy orders, and the probability of $C_{buy}^0$ and $C_{buy}^1$ increases.

$M_{buy}^0$: Increases the probability of $M_{buy}^0$. This can be explained by *order splitting*: large orders are split into smaller pieces that are more easily executed, and the *momentum effect*: other participants following the move. The increase of the probability of $M_{buy}^1$ and $L_{buy}^1$ is also explained by the momentum effect.

$L_{buy}^1$: Improves the offered price to buy the stock. The first major effect observed is a big increase in the probability of $M_{sell}^1$, which is precisely the market taking effect postulated in Section 4.3.2. The second effect is a large increase in the probability of $C_{buy}^1$ whereby the new liquidity is rapidly cancelled. This might reflect a market manipulation, where agents are posting fake orders.

$C_{buy}^1$: A total cancellation of the best buy limit increases the probability of $L_{buy}^1$; other participants re-offer the liquidity at the previous best buy price. It also increases the probability of $L_{sell}^1$, when a new consensus is concluded by the market participants at a lower price.

$M_{buy}^1$: Consumes all the offered liquidity at the best ask. This increases the probability of $L_{sell}^1$ when some participants re-offer the liquidity at the same previous best ask price. It also increases the probability of $L_{buy}^1$, when a new consensus is concluded by the market participants at a higher price. This is the market making effect already identified and studied in Section 4.3.1.

Much as these conditional probabilities are informative and practically useful, one must return to a finer description based on arrival times, as these are at the core of any microscopic description of a limit order book. The next section revisits the study of arrival times, distinguishing between aggressive and passive orders.

### 4.4.1 The fine structure of inter-event durations: Using lagged correlation matrices

In this section, we return to the study of inter-event durations as in Section 4.3.1, but specializing to all the event types we have introduced.

Table 4.4 represents the median of the waiting time (in seconds) to the event $j$ (in column) since, the last observed event $i$ (in row) and Table 4.5 represents the mean of this waiting time.

Beyond these basic measurements, some insight can be gained by studying the covariance matrix of these inter-arrival times.

**Table 4.4** Median conditional waiting time

| | $L^0_{buy}$ | $L^0_{sell}$ | $C^0_{buy}$ | $C^0_{sell}$ | $M^0_{buy}$ | $M^0_{sell}$ | $L^1_{buy}$ | $L^1_{sell}$ | $C^1_{buy}$ | $C^1_{sell}$ | $M^1_{buy}$ | $M^1_{sell}$ |
|---|---|---|---|---|---|---|---|---|---|---|---|---|
| $\|L^0_{buy}$ | 0.019 | 0.884 | 0.564 | 0.304 | 13.96 | 17.21 | 10.97 | 13.85 | 43.64 | 37.03 | 18.94 | 25.43 |
| $\|L^0_{sell}$ | 0.888 | 0.017 | 0.327 | 0.556 | 17.25 | 13.76 | 13.92 | 10.96 | 38.02 | 41.97 | 25.79 | 18.32 |
| $\|C^0_{buy}$ | 0.398 | 0.130 | 0.015 | 0.987 | 17.02 | 16.15 | 13.06 | 11.55 | 37.23 | 40.06 | 24.41 | 22.23 |
| $\|C^0_{sell}$ | 0.137 | 0.391 | 0.975 | 0.012 | 16.16 | 16.65 | 11.47 | 12.81 | 41.00 | 35.73 | 22.52 | 23.56 |
| $\|M^0_{buy}$ | 0.002 | 0.045 | 0.168 | 0.006 | 0.01 | 8.78 | 0.82 | 6.03 | 31.22 | 21.06 | 0.10 | 15.48 |
| $\|M^0_{sell}$ | 0.041 | 0.002 | 0.006 | 0.154 | 9.09 | 0.01 | 6.34 | 0.81 | 22.70 | 28.60 | 15.91 | 0.08 |
| $\|L^1_{buy}$ | 0.005 | 0.084 | 0.258 | 0.009 | 7.48 | 7.45 | 4.29 | 6.58 | 13.59 | 22.27 | 10.39 | 7.93 |
| $\|L^1_{sell}$ | 0.084 | 0.005 | 0.012 | 0.240 | 7.29 | 7.13 | 6.58 | 4.04 | 23.96 | 13.52 | 7.54 | 9.88 |
| $\|C^1_{buy}$ | 0.019 | 0.019 | 0.004 | 0.439 | 14.09 | 16.62 | 2.09 | 6.48 | 11.46 | 28.40 | 18.81 | 20.83 |
| $\|C^1_{sell}$ | 0.017 | 0.021 | 0.400 | 0.004 | 15.76 | 13.03 | 6.15 | 2.07 | 27.81 | 11.98 | 20.07 | 17.53 |
| $\|M^1_{buy}$ | 0.003 | 0.033 | 0.158 | 0.003 | 5.47 | 9.35 | 1.21 | 1.99 | 27.82 | 21.73 | 7.24 | 14.96 |
| $\|M^1_{sell}$ | 0.030 | 0.003 | 0.003 | 0.139 | 9.40 | 5.60 | 2.32 | 1.16 | 21.77 | 24.91 | 14.89 | 7.86 |

**Table 4.5** Mean conditional waiting time

| | $L^0_{buy}$ | $L^0_{sell}$ | $C^0_{buy}$ | $C^0_{sell}$ | $M^0_{buy}$ | $M^0_{sell}$ | $L^1_{buy}$ | $L^1_{sell}$ | $C^1_{buy}$ | $C^1_{sell}$ | $M^1_{buy}$ | $M^1_{sell}$ |
|---|---|---|---|---|---|---|---|---|---|---|---|---|
| $\|L^0_{buy}$ | 1.47 | 3.35 | 2.68 | 3.25 | 30.20 | 33.02 | 27.40 | 29.81 | 96.89 | 86.52 | 41.32 | 46.86 |
| $\|L^0_{sell}$ | 3.37 | 1.46 | 3.31 | 2.70 | 33.21 | 29.75 | 30.30 | 27.07 | 91.59 | 91.45 | 47.68 | 40.29 |
| $\|C^0_{buy}$ | 2.28 | 2.24 | 1.74 | 3.76 | 33.08 | 32.32 | 29.17 | 27.69 | 89.10 | 89.16 | 46.23 | 43.99 |
| $\|C^0_{sell}$ | 2.24 | 2.27 | 3.73 | 1.72 | 32.52 | 32.55 | 27.91 | 28.57 | 94.05 | 84.18 | 44.79 | 44.96 |
| $\|M^0_{buy}$ | 0.67 | 1.63 | 2.00 | 1.50 | 10.83 | 24.57 | 15.89 | 21.97 | 86.25 | 73.85 | 17.35 | 36.77 |
| $\|M^0_{sell}$ | 1.63 | 0.67 | 1.52 | 1.98 | 25.15 | 10.82 | 22.71 | 15.65 | 79.56 | 79.61 | 37.68 | 17.13 |
| $\|L^1_{buy}$ | 0.95 | 1.84 | 2.55 | 1.53 | 23.81 | 23.35 | 18.55 | 20.94 | 61.80 | 68.31 | 32.01 | 27.93 |
| $\|L^1_{sell}$ | 1.82 | 0.96 | 1.58 | 2.50 | 23.20 | 23.41 | 21.50 | 18.13 | 75.24 | 59.76 | 27.80 | 31.44 |
| $\|C^1_{buy}$ | 1.48 | 1.75 | 1.32 | 3.15 | 30.36 | 33.25 | 13.71 | 20.72 | 50.47 | 72.32 | 39.63 | 41.57 |
| $\|C^1_{sell}$ | 1.54 | 1.43 | 2.91 | 1.23 | 32.53 | 29.31 | 20.64 | 13.79 | 76.86 | 49.83 | 41.55 | 38.24 |
| $\|M^1_{buy}$ | 0.62 | 1.51 | 1.98 | 1.12 | 21.74 | 25.38 | 15.51 | 15.22 | 80.46 | 71.22 | 28.70 | 36.06 |
| $\|M^1_{sell}$ | 1.45 | 0.59 | 1.09 | 1.91 | 25.45 | 21.65 | 15.92 | 14.81 | 73.89 | 72.67 | 36.10 | 29.11 |

In general, consider a stationary $M$-dimensional *point process* $N$ with coordinates $(N^1(t), \ldots, N^M(t))$, $t \in \mathbb{R}_+$: $N^i(t)$ is the number of events of type $i$ having occured up to time $t$ (see Appendix C.1 for the definition and basic properties of point processes). Given a duration $h$ and a lag $\tau$, the *lagged covariance matrix* $C^h_\tau = \left(C^h_\tau(i,j)\right)_{1 \leq i,j \leq M}$ of the process is defined by:

$$C^h_\tau(i,j) = \frac{1}{h} \text{Cov}(N^i(t+h+\tau) - N^i(t+\tau), N^j(t+h) - N^j(t)). \quad (4.1)$$

When comparing two different types of events, such a construct helps identify dependencies, and possible *lead – lag* relationships, see for example, Bouchaud and Potters (2004), and Bacry et al. (2012) for a detailed study of this measure in the context of mutually excited point processes. Its decay as a function of the lag value $\tau$ is also informative on potential long-range interactions.

In order to avoid side effects caused by the wide variability of the frequencies across event type, it is actually more robust to rely on the lagged linear correlation matrix

$$\rho_\tau^h(i,j) = \text{Correlation}(N^i(t+h+\tau) - N^i(t+\tau), N^j(t+h) - N^j(t)). \tag{4.2}$$

In the case under scrutiny, the components of the process $N$ correspond to the events $M_{\text{buy}}^0, M_{\text{sell}}^0, M_{\text{buy}}^1, M_{\text{sell}}^1, L_{\text{buy}}^0, L_{\text{sell}}^0, L_{\text{buy}}^1, L_{\text{sell}}^1, C_{\text{buy}}^0, C_{\text{sell}}^0, C_{\text{buy}}^1, C_{\text{sell}}^1$ as in Table 4.1.

The time step $h$ is chosen as 0.1 second, and $\tau \in \{0.1, 0.2, \ldots 0.9\}$. The empirical lagged correlation coefficients are computed per day per stock. The results are then averaged across stocks and days. For each pair of event types $(i,j)$, the function $\tau \mapsto \rho_\tau^h(i,j)$ describes the temporal decay of the impact events of type $j$ have on events of type $i$. Compared with the rather crude approach based on conditional probabilities, it provides a finer, richer description of the temporal dependency structure between events.

Figure 4.5 details the impact of the different types of orders on the arrival of $M_{\text{buy}}^1$ orders.

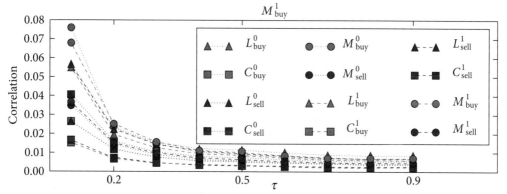

**Fig. 4.5** Impact functions on $M_{\text{buy}}^1$ arrival intensity: the graph confirms that the most relevant events to explain the instantaneous intensity of $M_{\text{buy}}^1$ are $M_{\text{buy}}^0, M_{\text{buy}}^1$ and $L_{\text{sell}}^1$

We see that the intensity of aggressive market orders $M_{\text{buy}}^1$ is primarily correlated with previous market orders on the same side ($M_{\text{buy}}^1$ and $M_{\text{buy}}^0$), which expresses the classical clustering phenomenon. More interestingly, the fact that it is also highly correlated with past aggressive limit orders is an evidence of the market taking effect. As already noted, this market taking effect does not necessarily occur on the same side ($L_{\text{sell}}^1$), but rather, affects both sides: $L_{\text{sell}}^1$ and $L_{\text{buy}}^1$ exhibit similar correlation levels with $M_{\text{buy}}^1$.

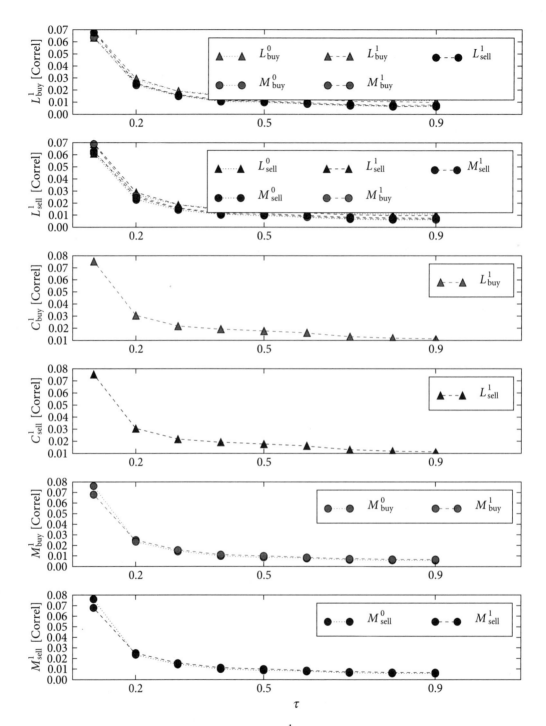

**Fig. 4.6** Impact functions on the six events $O^1$

For the sake of completeness, we provide in Fig. 4.6 the same results computed for the six types of aggressive events. In order to plot only the most relevant information, an arbitrary threshold of 6% is chosen: events for which the highest correlation coefficient is lower than 6% are discarded. Also, possible asymmetries in the lagged correlation curves between events of respective types $i$ and $j$ can be interpreted in terms of lead-lag relationships.

The intensity of $L^1_{\text{buy}}$ event is increased by the arrival of any $L_{\text{buy}}$ or $M_{\text{buy}}$ event. This means that liquidity providers follow on average the market consensus and provide more aggressive prices when the stock seems to move in the expected direction. Cancellations show a primary correlation with limit orders on the same side. This corresponds to the numerous observations where a new limit is rapidly cancelled. These results are clearly in line with the conclusions of Section 4.4 based on the conditional probabilities.

As a final comment, Fig. 4.6 clearly shows that the bid and ask sides exhibit symmetric behaviours with respect to all order types. This justifies the fact that many empirical studies on limit order books merge samples from the bid and ask sides, as was done in Section 4.3.1.

## 4.5 Conclusion

This chapter has provided numerous empirical evidences of the *clustering* of orders, of a *market making* (or: Liquidity resilience) effect and also of a *market taking* effect (market participants tends to seize the liquidity when it tightens the spread below usual levels). Different methods have been presented, relying either on inter-arrival times or on conditional probabilites of occurrence. These results lay the ground for the "better order book models" that we present in Chapter 8.

# PART TWO
# MATHEMATICAL MODELLING OF LIMIT ORDER BOOKS

# CHAPTER 5

# Agent-based Modelling of Limit Order Books: A Survey

## 5.1 Introduction

This chapter is dedicated to a review of *agent-based* models of limit order books, that is, models depicting, at the individual agent level, possibly from a statistical point of view, the interactions that lead to a transaction on a financial market. Far from being exhaustive, the survey is based on selected models that we feel are representative of some important, specific trends in agent-based modelling.

Although known, at least partly, for a long time – Mandelbrot (1963) gives a reference for a paper dealing with non-normality of price time series in 1915, followed by several others in the 1920's – some stylized facts of asset returns (heavy tails, volatility clustering, etc.) have often been left aside when modelling financial markets. They were even often referred to as "anomalous" characteristics, as if observations failed to comply with theory. Much has been done these past twenty years in order to address this challenge and provide new models that can reproduce these facts. These recent developments have been built on top of early attempts at modelling mechanisms of financial markets with agents. Stigler (1964), investigating some rules of the SEC[1], or Garman (1976), investigating double-auction microstructure, belong to those historical works. It seems that the first modern attempts at that type of models were made in the field of behavioural finance. This field aims at improving financial modelling based on the psychology and sociology of the investors. Models are built with agents who can exchange shares of stocks according to exogenously defined utility functions reflecting their preferences and risk aversions. LeBaron (2006a,b) shows that this type of modelling offers good flexibility for reproducing some of the stylized facts and provides a review of

---
[1] Securities and Exchange Commission, the agency supervising the organization of regulator of the US stock exchanges

that type of model. However, although achieving some of their goals, these models suffer from many drawbacks: First, they are very complex, and it may be a very difficult task to identify the role of their numerous parameters and the types of dependence on these parameters; second, the chosen utility functions do not necessarily reflect what is observed on the mechanisms of a financial market.

A sensible change in modelling appears with much simpler models implementing only well-identified and presumably realistic "behaviour": Cont and Bouchaud (2000) uses noise traders that are subject to "herding", i.e., formation of random clusters of traders sharing the same view on the market. The idea is used in Raberto et al. (2001) as well. A complementary approach is to characterize traders as fundamentalists, chartists or noise traders. Lux and Marchesi (2000) propose an agent-based model in which these types of traders interact. In all these models, the price variation directly results from the excess demand: at each time step, all agents submit orders and the resulting price is computed. Therefore, everything is cleared at each time step and there is no structure of order book to keep track of orders.

One big step is made with models really taking into account limit orders and keeping them in an order book once submitted and not executed.Chiarella and Iori (2002) build an agent-based model where all traders submit orders depending on the three elements identified in Lux and Marchesi (2000): Chartists, fundamentalists, noise. Orders submitted are then stored in a persistent order book. In fact, one of the first simple models with this feature was proposed in Bak et al. (1997). In this model, orders are particles moving along a price line, and each collision is a transaction. Due to numerous caveats in this model, the authors propose in the same paper an extension with fundamentalist and noise traders in the spirit of the models previously evoked. Maslov (2000) goes further in the modelling of trading mechanisms by taking into account fixed limit orders and market orders that trigger transactions, and really simulating the order book. This model was analytically solved using a mean-field approximation by Slanina (2001).

Following this trend of modelling, the more or less "rational" agents composing models in Economics tends to vanish and be replaced by the notion of flows: orders are not submitted any more by an agent following a strategic behaviour, but are viewed as an arriving flow whose properties are to be determined by empirical observations of market mechanisms. Challet and Stinchcombe (2001) propose a simple model of order flows: limit orders are deposited in the order book and can be removed if not executed, in a simple deposition-evaporation process. Bouchaud et al. (2002) use this type of model with empirical distribution as inputs. Mike and Farmer (2008) is a very complete empirical model, where order placement and cancellation models are proposed and fitted on empirical data. Finally, new challenges arise as scientists try to identify simple mechanisms that allow an agent-based model to reproduce non-trivial behaviours:

Herding behaviour in Cont and Bouchaud (2000), dynamic price placement in Preis et al. (2007), threshold behaviour in Cont (2007), etc.

## 5.2 Early Order-driven Market Modelling: Market Microstructure and Policy Issues

The pioneering works in simulation of financial markets were aimed to study market regulations. The very first one, Stigler (1964), tries to investigate the effect of regulations of the SEC on American stock markets, using empirical data from the 1920s and the 1950s. Twenty years later, at the start of the computerization of financial markets, Hakansson et al. (1985) implements a simulator in order to test the feasibility of automated market making. We will not review the huge microstructure literature in the line of the books by O'Hara (1997) or Hasbrouck (2007). However, by presenting a small selection of early models, we underline here the grounding of recent order book modelling.

### 5.2.1 A pioneer order book model

To our knowledge, the first attempt to simulate a financial market was by Stigler (1964). This paper was a biting and controversial reaction to the Report of the Special Study of the Securities Markets of the SEC [Cohen (1963b)], whose aim was to "study the adequacy of rules of the exchange and that the New York stock exchange undertakes to regulate its members in all of their activities" [Cohen (1963a)]. According to Stigler, this SEC report lacks rigorous tests when investigating the effects of regulation on financial markets. Stating that "demand and supply are [...] erratic flows with sequences of bids and asks dependent upon the random circumstances of individual traders", he proposes a simple simulation model to investigate the evolution of the market. In this model, constrained by simulation capability in 1964, price is constrained within $L = 10$ ticks. (Limit) orders are randomly drawn, in trade time, as follows: they can be bid or ask orders with equal probability, and their price level is uniformly distributed on the price grid. Each time an order crosses the opposite best quote, it is a market order. All orders are of size one. Orders not executed $N = 25$ time steps after their submission are cancelled. Thus, $N$ is the maximum number of orders available in the order book.

In the original paper, a run of a hundred trades was manually computed using tables of random numbers. Of course, no particular results concerning the stylized facts of financial time series was expected at that time. However, in his review of some order book models, Slanina (2008) makes simulations of a similar model, with parameters $L = 5000$ and $N = 5000$, and shows that price returns are not Gaussian: Their distribution exhibits power law with exponent 0.3, far from empirical data. As expected, the limitation $L$ is responsible for a sharp cut-off of the tails of this distribution.

### 5.2.2 Microstructure of the double auction

Garman (1976) provides an early study of the double auction market with a point of view that does not ignore temporal structure, and really defines order flows. Price is discrete and constrained to be within $\{p_1, ..., p_L\}$. Buy and sell orders are assumed to be submitted according to two Poisson processes of intensities $\lambda$ and $\mu$. Each time an order crosses the best opposite quote, it is a market order. All quantities are assumed to be equal to one. The aim of the author was to provide an empirical study of the market microstructure. The main result of its Poisson model was to support the idea that negative correlation of consecutive price changes is linked to the microstructure of the double auction exchange. This paper is very interesting because it can be seen as a precursor that clearly establishes the challenges of order book modelling. First, the mathematical formulation is promising. With its fixed constrained prices, Garman (1976) can define the state of the order book at a given time as the vector $(n_i)_{i=1,...,L}$ of awaiting orders (negative quantity for bid orders, positive for ask orders). Future analytical models will use similar vector formulations that can be cast into known mathematical processes in order to extract analytical results – see for example, Cont et al. (2010) reviewed below. Second, the author points out that although the Poisson model is simple, analytical solution is hard to work out, and he then provides Monte Carlo simulation. The need for numerical and empirical developments is a constant in all following models. Third, the structural question is clearly asked in the conclusion of the paper: "Does the auction-market model imply the characteristic leptokurtosis seen in empirical security price changes?". The computerization of markets that was about to take place when this research was published – Toronto's CATS[2] opened a year later in 1977 – motivated many following papers on the subject.

### 5.2.3 Zero-intelligence

In the models by Stigler (1964) and Garman (1976), orders are submitted in a purely random way on the grid of possible prices. Traders do not observe the market here and do not act according to a given strategy. Thus, these two contributions clearly belong to a class of "zero-intelligence" models. Gode and Sunder (1993) is (one of) the first papers to introduce the expression "zero-intelligence" in order to describe non-strategic behaviour of traders. It is applied to traders that submit random orders in a double auction market. The expression has since been widely used in agent-based modelling, sometimes in a slightly different meaning (see more recent models described in this review). In Gode and Sunder (1993), two types of zero-intelligence traders are studied. The first are unconstrained zero-intelligence traders. These agents can submit random order at random prices, within the allowed price range $\{1, ..., L\}$. The second are constrained zero-intelligence traders. These agents submit random orders as well, but with the

---
[2] Computer Assisted Trading System

constraint that they cannot cross their given reference price $p_i^R$: Constrained zero-intelligence traders are not allowed to buy or sell at loss. The aim of the authors was to show that double auction markets exhibit an intrinsic "allocative efficiency" (ratio between the total profit earned by the traders divided by the maximum possible profit) even with zero-intelligence traders. An interesting fact is that in this experiment, price series resulting from actions by zero-intelligence traders are much more volatile than the ones obtained with constrained traders. This fact will be confirmed in future models where "fundamentalists" traders, having a reference price, are expected to stabilize the market (see Wyart and Bouchaud (2007) or Lux and Marchesi (2000) below). Note that the results have been criticized by Cliff and Bruten (1997), who show that the observed convergence of the simulated price towards the theoretical equilibrium price may be an artefact of the model. More precisely, the choice of traders' demand carry a lot of constraints that alone explain the observed results.

Modern works in econophysics owe a lot to these early models or contributions. Starting in the mid-90s, physicists have proposed simple order book models directly inspired from physics, where the analogy "order ≡ particle" is emphasized. Three main contributions are presented in the next section.

## 5.3 Order-driven Market Modelling in Econophysics

### 5.3.1 The order book as a reaction-diffusion model

A very simple model directly taken from physics was presented in Bak et al. (1997). The authors consider a market with $N$ noise traders able to exchange one share of stock at a time. Price $p(t)$ at time $t$ is constrained to be an integer (i.e. price is quoted in number of ticks) with an upper bound $\bar{p}$: $\forall t, p(t) \in \{0, \ldots, \bar{p}\}$. Simulation is initiated at time 0 with half of the agents asking for one share of stock (buy orders, bid) with price:

$$p_b^j(0) \in \{0, \bar{p}/2\}, \quad j = 1, \ldots, N/2, \tag{5.1}$$

and the other half offering one share of stock (sell orders, ask) with price:

$$p_s^j(0) \in \{\bar{p}/2, \bar{p}\}, \quad j = 1, \ldots, N/2. \tag{5.2}$$

At each time step $t$, agents revise their offer by exactly one tick, with equal probability to go up or down. Therefore, at time $t$, each seller (resp. buyer) agent chooses his new price as:

$$p_s^j(t+1) = p_s^j(t) \pm 1 \quad (\text{resp. } p_b^j(t+1) = p_b^j(t) \pm 1). \tag{5.3}$$

A transaction occurs when there exists $(i,j) \in \{1,\ldots,N/2\}^2$ such that $p_b^i(t+1) = p_s^j(t+1)$. In such a case the orders are removed and the transaction price is recorded as the new price $p(t)$. Once a transaction has been recorded, two orders are placed at the extreme positions on the grid: $p_b^i(t+1) = 0$ and $p_s^j(t+1) = \bar{p}$. As a consequence, the number of orders in the order book remains constant and equal to the number of agents. In Fig. 5.1, an illustration of these moving particles is given.

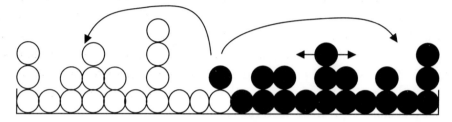

**Fig. 5.1** Illustration of the Bak, Paczuski and Shubik model: White particles (buy orders, bid) moving from the left, black particles (sell orders, ask) moving from the right. Reproduced from Bak et al. (1997)

**Table 5.1** Analogy between the $A + B \rightarrow \emptyset$ reaction model and the order book in Bak et al. (1997)

| Physics | Bak et al. (1997) |
|---|---|
| Particles | Orders |
| Finite Pipe | Order book |
| Collision | Transaction |

Following this analogy, it can thus be shown that the variation $\Delta p(t)$ of the price $p(t)$ verifies:

$$\Delta p(t) \sim t^{1/4} \left( \ln\left(\frac{t}{t_0}\right) \right)^{1/2}. \tag{5.4}$$

Thus, at long time scales, the series of price increments simulated in this model exhibit a Hurst exponent $H = 1/4$. As for the stylized fact $H \approx 0.7$, this sub-diffusive behaviour appears to be a step in the wrong direction compared to the random walk $H = 1/2$. Moreover, Slanina (2008) points out that no fat tails are observed in the distribution of the returns of the model, but rather fits the empirical distribution with an exponential decay. Other drawbacks of the model could be mentioned. For example, the reintroduction of orders at each end of the pipe leads to unrealistic shape of the order book, as shown on Fig. 5.2.

Actually here is the main drawback of the model: "Moving" orders is highly unrealistic as for modelling an order book, and since, it does not reproduce any known financial exchange mechanism, it cannot be the base for any larger model.

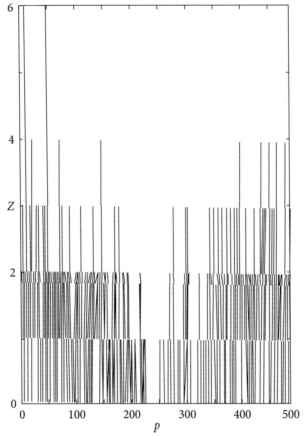

**Fig. 5.2** Snapshot of the limit order book in the Bak, Paczuski and Shubik model. Reproduced from Bak et al. (1997)

Therefore, attempts by the authors to build several extensions of this simple framework, in order to reproduce stylized facts by adding fundamental traders, strategies, trends, etc. are not of interest for us in this review. However, we feel that the basic model as such is very interesting because of its simplicity and its "particle" representation of an order-driven market that has opened the way for more realistic models.

### 5.3.2 Introducing market orders

Maslov (2000) keeps the zero-intelligence structure of the Bak et al. (1997) model but adds more realistic features in the order placement and evolution of the market. First, limit orders are submitted and stored in the model, without moving. Second, limit orders are submitted around the best quotes. Third, market orders are submitted to trigger transactions. More precisely, at each time step, a trader is chosen to perform an action. This trader can either submit a limit order with probability $q_l$ or submit a market order with probability $1 - q_l$. Once this choice is made, the order is a buy or sell order with equal probability. All orders have a one unit volume.

As usual, we denote $p(t)$ the current price. In case the submitted order at time step $t+1$ is a limit ask (resp. bid) order, it is placed in the book at price $p(t) + \Delta$ (resp. $p(t) - \Delta$), $\Delta$ being a random variable uniformly distributed in $[0; \Delta^M = 4]$. In case the submitted order at time step $t+1$ is a market order, one order at the opposite best quote is removed and the price $p(t+1)$ is recorded. In order to prevent the number of orders in the order book from large increase, two mechanisms are proposed by the author: Either keeping a fixed maximum number of orders (by discarding new limit orders when this maximum is reached), or removing them after a fixed lifetime if they have not been executed.

Numerical simulations show that this model exhibits non-Gaussian heavy-tailed distributions of returns. In Fig. 5.3, the empirical probability density of the price increments for several time scales are plotted.

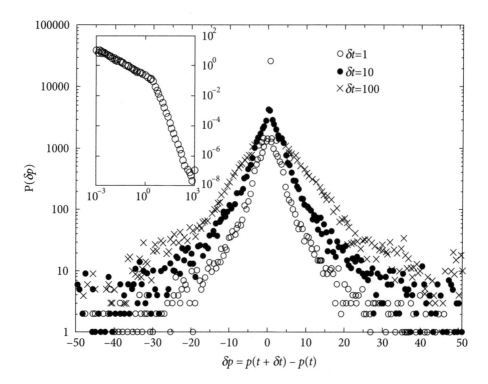

**Fig. 5.3** Empirical probability density functions of the price increments in the Maslov model. In inset, log-log plot of the positive increments. Reproduced from Maslov (2000)

For a time scale $\delta t = 1$, the author fit the tails distribution with a power law with exponent 3.0, i.e. reasonable compared to empirical value. However, the Hurst exponent of the price series is still $H = 1/4$ with this model. It should also be noted that Slanina (2001) proposed an analytical study of the model using a mean-field approximation.

This model brings very interesting innovations in order book simulation: order book with (fixed) limit orders, market orders, necessity to cancel orders waiting too long in the order book. These features are of prime importance in any following order book model.

### 5.3.3 The order book as a deposition-evaporation process

Challet and Stinchcombe (2001) continue the work of Bak et al. (1997) and Maslov (2000), and develop the analogy between dynamics of an order book and an infinite one-dimensional grid, where particles of two types (ask and bid) are subject to three types of events: *deposition* (limit orders), *annihilation* (market orders) and *evaporation* (cancellation). Note that annihilation occurs when a particle is deposited on a site occupied by a particle of another type. The analogy is summarized in Table 5.2.

**Table 5.2** Analogy between the deposition-evaporation process and the order book in Challet and Stinchcombe (2001)

| Physics | Challet and Stinchcombe (2001) |
|---|---|
| Particles | Orders |
| Infinite lattice | Order book |
| Deposition | Limit orders submission |
| Evaporation | Limit orders cancellation |
| Annihilation | Transaction |

Hence, the model goes as follows: at each time step, a bid (resp. ask) order is deposited with probability $\lambda$ at a price $n(t)$ drawn according to a Gaussian distribution centred on the best ask $a(t)$ (resp. best bid $b(t)$) and with variance depending linearly on the spread $s(t) = a(t) - b(t)$: $\sigma(t) = Ks(t) + C$. If $n(t) > a(t)$ (resp. $n(t) < b(t)$), then it is a market order: annihilation takes place and the price is recorded. Otherwise, it is a limit order and it is stored in the book. Finally, each limit order stored in the book has a probability $\delta$ to be cancelled (evaporation).

Figure 5.4 shows the average return as a function of the time scale. It appears that the series of price returns simulated with this model exhibit a Hurst exponent $H = 1/4$ for short time scales, and that tends to $H = 1/2$ for larger time scales. This behaviour might be the consequence of the random evaporation process [which was not modelled in Maslov (2000), where $H = 1/4$ for large time scales]. Although some modifications of the process (more than one order per time step) seem to shorten the sub-diffusive region, it is clear that no over-diffusive behaviour is observed.

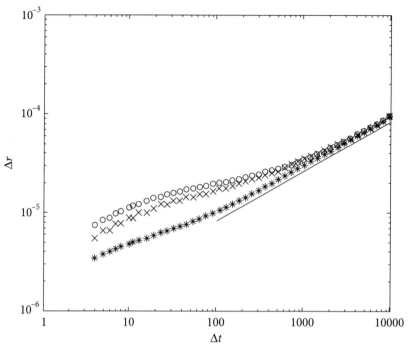

**Fig. 5.4** Average return $\langle r_{\Delta t} \rangle$ as a function of $\Delta t$ for different sets of parameters and simultaneous depositions allowed in the Challet and Stinchcombe model. Reproduced from Challet and Stinchcombe (2001)

## 5.4 Empirical Zero-intelligence Models

The three models presented in Section 5.3 have successively isolated essential mechanisms that are to be used when simulating a "realistic" market: One order is the smallest entity of the model; the submission of one order is the time dimension (i.e. event time is used, not an exogenous time defined by market clearing and "tatonnement" on exogenous supply and demand functions); submission of market orders [as such in Maslov (2000), as "crossing limit orders" in Challet and Stinchcombe (2001)] and cancellation of orders are taken into account. On the one hand, one may try to describe these mechanisms using a small number of parameters, using Poisson process with constant rates for order flows, constant volumes, etc. This might lead to some analytically tractable models, such as the ones presented in Chapters 6 and 7. On the other hand, one may try to fit more complex empirical distributions to market data without analytical concern.

The latter type of modelling is best represented by Mike and Farmer (2008). It is the first model that proposes an advanced calibration on the market data as for order placement and cancellation methods. As for volume and time of arrivals, assumptions of previous models still hold: all orders have the same volume, discrete event time is used for simulation, i.e.

one order (limit or market) is submitted per time step. Following Challet and Stinchcombe (2001), there is no distinction between market and limit orders, i.e. market orders are limit orders that are submitted across the spread $s(t)$. More precisely, at each time step, one trading order is simulated: An ask (resp. bid) trading order is randomly placed at $n(t) = a(t) + \delta a$ (resp. $n(t) = b(t) + \delta b$) according to a Student distribution with scale and degrees of freedom calibrated on market data. If an ask (resp. bid) order satisfies $\delta a < -s(t) = b(t) - a(t)$ (resp. $\delta b > s(t) = a(t) - b(t)$), then it is a buy (resp. sell) market order and a transaction occurs at price $a(t)$ (resp. $b(t)$).

During a time step, several cancellations of orders may occur. The authors propose an empirical distribution for cancellation based on three components for a given order:

- The position in the order book, measured as the ratio $y(t) = \frac{\Delta(t)}{\Delta(0)}$ where $\Delta(t)$ is the distance of the order from the opposite best quote at time $t$,
- The order book imbalance, measured by the indicator $N_{\text{imb}}(t) = \frac{N_a(t)}{N_a(t)+N_b(t)}$ (resp. $N_{\text{imb}}(t) = \frac{N_b(t)}{N_a(t)+N_b(t)}$) for ask (resp. bid) orders, where $N_a(t)$ and $N_b(t)$ are the number of orders at ask and bid in the book at time $t$,
- The total number $N(t) = N_a(t) + N_b(t)$ of orders in the book.

Their empirical study leads them to assume that the cancellation probability has an exponential dependence on $y(t)$, a linear one in $N_{\text{imb}}$ and finally decreases approximately as $1/N_t(t)$ as for the total number of orders. Thus, the probability $P(C|y(t), N_{\text{imb}}(t), N_t(t))$ to cancel an ask order at time $t$ is formally written:

$$P(C|y(t), N_{\text{imb}}(t), N_t(t)) = A(1 - e^{-y(t)})(N_{\text{imb}}(t) + B)\frac{1}{N_t(t)}, \qquad (5.5)$$

where the constants $A$ and $B$ are to be fitted on market data. Figure 5.5 shows that this empirical formula provides a quite good fit on market data.

Finally, the authors mimic the observed long memory of order signs by simulating a fractional Brownian motion. The auto-covariance function $\Gamma(t)$ of the increments of such a process exhibits a slow decay:

$$\Gamma(k) \sim H(2H-1)t^{2H-2} \qquad (5.6)$$

and it is therefore easy to reproduce exponent $\beta$ of the decay of the empirical autocorrelation function of order signs observed on the market with $H = 1 - \beta/2$.

The results of this empirical model are quite satisfying as for return and spread distribution. The distribution of returns exhibit fat tails which are in agreement with empirical data, as shown in Fig. 5.6.

## 56  Limit Order Books

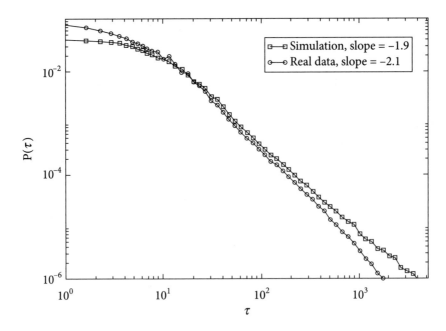

**Fig. 5.5**  Lifetime of orders for simulated data in the Mike and Farmer model, compared to the empirical data used for fitting. Reproduced from Mike and Farmer (2008)

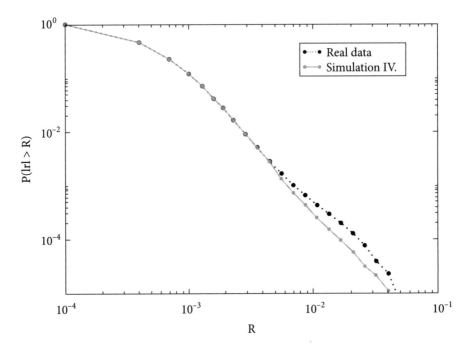

**Fig. 5.6**  Cumulative distribution of returns in the Mike and Farmer model, compared to the empirical data used for fitting. Reproduced from Mike and Farmer (2008)

The spread distribution is also very well reproduced. As their empirical model has been built on the data of only one stock, the authors test their model on 24 other data sets of stocks on the same market and find for half of them a good agreement between empirical and simulated properties. However, the bad results of the other half suggest that such a model is still far from being "universal".

Despite these very nice results, some drawbacks have to be pointed out. The first one is the fact that the stability of the simulated order book is far from ensured. Simulations using empirical parameters in the simulations may bring situations where the order book is emptied by large consecutive market orders. Thus, the authors require that there be at least two orders on each side of the book. This exogenous trick might be important, since it is activated precisely in the case of rare events that influence the tails of the distributions. Also, the original model does not focus on volatility clustering. Gu and Zhou (2009) propose a variant that tackles this feature. Another important drawback of the model is the way order signs are simulated. As noted by the authors, using an exogenous fractional Brownian motion leads to correlated price returns, which is in contradiction with empirical stylized facts. We also find that at long time scales it leads to a dramatic increase of volatility. As already seen in Chapter 2, the correlation of trade signs can be at least partly seen as an artefact of execution strategies. Therefore this element is one of the numerous that should be taken into account when "programming" the agents of the model. In order to do so, we have to leave the (quasi) "zero-intelligence" world and see how modelling based on heterogeneous agents might help to reproduce non-trivial behaviours. Prior to presenting such developments in Chapter 8, we briefly review some analytical works on the "zero-intelligence" models.

## 5.5 Some Analytical and Mathematical Developments in Zero-intelligence Order Book Modelling

This last section is a brief introduction to the developments of Chapters 6 and 7.

Smith et al. (2003) investigates the scaling properties of some liquidity and price characteristics in a limit order book model. These results are summarized in Table 5.3. In Smith et al. (2003), orders arrive on an *infinite* price grid (This is consistent as the limit orders arrival rate *per price level* is finite). Moreover, the arrival rates are independent of the price level, which has the advantage of enabling the analytical predictions summarized in Table 5.3.

These results are obtained by mean-field approximations, which assume that the fluctuations at adjacent price levels are independent. This allows fruitful simplifications of the complex dynamics of the order book. In addition, the authors do not characterize the convergence of the coarse-grained price process in the sense of stochastic-process limits, nor do they show that the limiting process is precisely a Wiener process (Theorem 6.5).

**Table 5.3** Results of Smith et al. $\epsilon := q/\left(\lambda^M/2\lambda^C\right)$ is a "granularity" parameter that characterizes the effect of discreteness in order sizes, $p_c := \lambda^M/2\lambda^L$ is a characteristic price interval, and $f$ and $g$ are slowly varying functions

| Quantity | Scaling relation |
|---|---|
| Average asymptotic depth | $\lambda^L/\lambda^C$ |
| Average spread | $\lambda^M/\lambda^L f(\epsilon, \Delta P/p_c)$ |
| Slope of average depth profile | $\left(\lambda^L\right)^2/\lambda^M \lambda^C g(\epsilon, \Delta P/p_c)$ |
| Price "diffusion" parameter at short time scales | $\left(\lambda^M\right)^2 \lambda^C/\lambda^L \epsilon^{-0.5}$ |
| Price "diffusion" parameter at long time scales | $\left(\lambda^M\right)^2 \lambda^C/\lambda^L \epsilon^{0.5}$ |

Cont et al. (2010) is an important step in zero-intelligence modelling of limit order books, because of the simplicity of the model and the consequent analytical tractability. In this model, the bid and ask prices are integers, constrained on a grid $\{1, \ldots, n\}$ (in ticks). Limit orders are submitted according to Poisson processes with rate $\lambda_i$, where $i$ is the distance to the opposite best price. $i \geq 0$ ensures that these limit orders are not marketable. Market orders are also Poisson, with rate $\mu$. The rate of cancellations at a given price is proportional to the volume standing in the book at that price, which is equivalent to assume that each standing order can be cancelled after an exponential life time with parameter $\theta > 0$. All orders are unit-sized. This model is (one of) the first to clearly treat the order book as a complex Markovian queueing system. The authors show that this model admits a stationary state, propose some simulations and prove that some analytical results for quantities of interest can be obtained, such as the probability of increase of the mid-price, or the probability of execution before a mid-price move.

## 5.6 Conclusion

Several pioneering works in agent-based modelling of order-driven markets have been reviewed in this chapter, and the emphasis has been set on the more statistical approaches. Building on these earlier works, we will continue in Chapters 6, 8 and 9 our study of limit order book models by introducing a proper mathematical and numerical framework. Questions such as the ergodicity of the order book and invariance principles for the suitably rescaled price process will be studied, and numerical analyses will be performed.

# CHAPTER 6

# The Mathematical Structure of Zero-intelligence Models

## 6.1 Introduction

In this chapter, we introduce a general framework to study the mathematical properties of limit order books in a Markovian context. One of our main motivations is to understand the interplay between the structure of limit order books and more traditional objects of interest on financial markets, namely, the price and spread dynamics. After casting the study of limit order books in the appropriate setting of Markovian point processes, we derive several mathematical results in the case of independent Poissonian arrival times. In particular, we show that the cancellation rate plays an important role, ensuring the ergodicity of the order book and the exponential convergence towards its stationary distribution. We also address the convergence of the price process induced by the order book dynamics to a diffusive process at macroscopic time scales. This natural question has attracted a lot of interest of late, as it is an important building block in establishing the compatibility of microstructural models with more classical models used in continuous time finance.

### 6.1.1 An elementary approximation: Perfect market making

We start with the simplest agent-based market model, which we call the *Bachelier* market since it provides an order-driven representation of the Bachelier model for asset prices:

- The order book starts in a full state: All limits above $P^A(0)$ and below $P^B(0)$ are filled with one limit order of unit size $q$. The spread starts equal to 1 tick;
- The flow of market orders is modeled by two independent Poisson processes $M^+(t)$ (buy orders) and $M^-(t)$ (sell orders) with constant arrival rates (or intensities) $\lambda^+$ and $\lambda^-$ ;

- There is one liquidity provider, who reacts *immediately* after a market order arrives so as to maintain the spread constantly equal to 1 tick. He places a limit order on the same side as the market order (i.e. a buy limit order after a buy market order and vice versa) with probability $u$ and on the opposite side with probability $1 - u$.

The mid-price dynamics can be written in the following form

$$dP(t) = \Delta P \, (dM^+(t) - dM^-(t))Z, \tag{6.1}$$

where $Z$ is a Bernoulli random variable

$$Z = 0 \text{ with probability } (1 - u), \tag{6.2}$$

and

$$Z = 1 \text{ with probability } u, \tag{6.3}$$

independent on $M^+$ and $M^-$, and the price increment $\Delta P$ is equal to the tick size.

The infinitesimal generator $\mathcal{L}$ associated with this dynamics is

$$\mathcal{L}f(P) = u \left[ \lambda^+ \left( f(P + \Delta P) - f \right) + \lambda^- \left( f(P - \Delta P) - f \right) \right], \tag{6.4}$$

where $f$ denotes a test function. It is well known that a continuous limit is obtained under suitable assumptions on the intensity and tick size. Noting that Eq. (6.4) can be rewritten as

$$\begin{aligned}\mathcal{L}f(P) &= \frac{1}{2}u \left( \lambda^+ + \lambda^- \right) (\Delta P)^2 \frac{f(P + \Delta P) - 2f + f(P - \Delta P)}{(\Delta P)^2} \\ &\quad + u \left( \lambda^+ - \lambda^- \right) \Delta P \frac{f(P + \Delta P) - f(P - \Delta P)}{2 \Delta P},\end{aligned} \tag{6.5}$$

and under the following assumptions

$$u \left( \lambda^+ + \lambda^- \right) (\Delta P)^2 \longrightarrow \sigma^2 \text{ as } \Delta P \to 0, \tag{6.6}$$

and

$$u \left( \lambda^+ - \lambda^- \right) \Delta P \longrightarrow \mu \text{ as } \Delta P \to 0, \tag{6.7}$$

the generator converges to the classical diffusion operator

$$\frac{\sigma^2}{2} \frac{\partial^2 f}{\partial P^2} + \mu \frac{\partial f}{\partial P}, \tag{6.8}$$

corresponding to a Wiener process with drift. This simple case is worked out as an example of the type of limit theorems that we will be interested in in the sequel. Ofcourse, an alternate approach using the Functional Central limit Theorem (see Theorem C.8 in Appendix C.2.2), yields a similar results: For given, fixed values of $\lambda^+$, $\lambda^-$ and $\Delta P$, the rescaled-centred price process

$$\frac{P(nt) - n\mu t}{\sigma \sqrt{n}} \tag{6.9}$$

converges as $n \to \infty$, to a standard Wiener process, where

$$\sigma = \Delta P \sqrt{(\lambda^+ + \lambda^-) u}, \tag{6.10}$$

and

$$\mu = \Delta P (\lambda^+ - \lambda^-) u. \tag{6.11}$$

One can easily achieve more complex diffusive limits such as a local volatility model, see e.g., Ethier and Kurtz (2005) Theorem 4.1, by imposing that the limit is a function of $P$ and $t$:

$$u (\lambda^+ + \lambda^-)(\Delta P)^2 \to \sigma^2(P, t) \tag{6.12}$$

and

$$u (\lambda^+ - \lambda^-) \Delta P \to \mu(P, t). \tag{6.13}$$

This may indeed be the case if the original intensities are functions of $P$ and $t$ themselves.

## 6.2 Order Book Dynamics

In this section, we introduce the general setup and notations for the study of limit order books.

### 6.2.1 Model setup: Poissonian arrivals, reference frame and boundary conditions

**Order book representation**

Each side of the order book is supposed to be fully described by a *finite* number of limits $K$, ranging from 1 to $K$ ticks away from the best available opposite quote. We use the notation

$$(\mathbf{a}(t); \mathbf{b}(t)) := (a_1(t), \ldots, a_K(t); b_1(t), \ldots, b_K(t)),$$

where $\mathbf{a} := (a_1, \ldots, a_K)$ represents the ask side of the order book, $a_i$ being the number of shares available $i$ ticks away from the best opposite quote; and similarly for $\mathbf{b} := (b_1, \ldots, b_K)$ on the bid side. In other words, we adopt a *finite moving frame*. This representation reflects faithfully the limit order books as seen by traders on their screens. For this reason, $\mathbf{a}, \mathbf{b}$ will sometimes be referred to as the *visible* limits. Note that this representation is different from the ones used in Cont et al. (2010) or Smith et al. (2003) (see also Gatheral and Oomen (2010) for an interesting discussion).

The quantities $a_i, b_i$'s are supposed to live in the discrete space $q\mathbb{Z}$, where $q \in \mathbb{N}^*$ is the minimum order size on each specific market (the *lot size*), but most of our results can be extended to the more general case of real-valued $a_i, b_i$s when orders arrive with random sizes.

Of interest are also the *integrated* quantities, introduced in Chapter 3, that describe the shape of the order book: They are the cumulative depth $\mathbf{A}, \mathbf{B}$ defined by

$$\mathbf{A}_i := \sum_{k=1}^{i} a_k, \tag{6.14}$$

and

$$\mathbf{B}_i := \sum_{k=1}^{i} |b_k|. \tag{6.15}$$

Also useful are the generalized inverse functions thereof

$$\mathbf{A}^{-1}(q') := \inf\{p : \mathbf{A}_p > q'\}, \tag{6.16}$$

and

$$\mathbf{B}^{-1}(q') := \inf\{p : \mathbf{B}_p > q'\}, \tag{6.17}$$

where $q'$ designates a certain quantity of shares. For the sake of notational simplicity, one can conveniently introduce the indices corresponding to the first non-empty limit. Their common value, which will be denoted by $i_S$, is equal to the spread $S$ in *number of ticks*:

$$i_S := \mathbf{A}^{-1}(0) = \mathbf{B}^{-1}(0) = \frac{S}{\Delta P}. \tag{6.18}$$

The boundary conditions described below will ensure that $i_S < \infty$.

## Boundary conditions

Constant boundary conditions are imposed outside the moving frame of size $2K$: every time the moving frame leaves a price level, the number of shares at that level is set to $a_\infty$ or $b_\infty$, depending on the side of the book. Our choice of a finite moving frame and constant boundary conditions has three motivations: Firstly, it assures that the order book does not become empty and that the best ask (resp. best bid) price $P^A$ (resp. $P^B$) is always defined. Secondly, it keeps the spread $S$ and the increments of $P^A$, $P^B$ bounded - this will be important when addressing the scaling limit of the price. Thirdly, it helps make the order book model Markovian, as we do not keep track of the price levels that have been visited, and then left, by the moving frame at some prior time.

Figure 6.1 is a representation of the order book using the above notations.

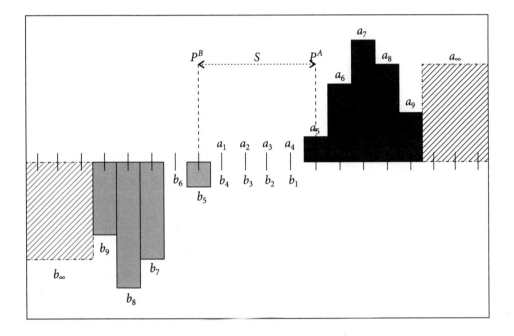

**Fig. 6.1** Order book dynamics: In this example, $K = 9$, $q = 1$, $a_\infty = 4$, $b_\infty = -4$. The shape of the order book is such that $\mathbf{a} = (0, 0, 0, 0, 1, 3, 5, 4, 2)$ and $\mathbf{b} = (0, 0, 0, 0, -1, 0, -4, -5, -3)$. The spread in ticks is given by $i_S = 5$. Assume that a sell market order arrives, then $\mathbf{a}, \mathbf{b}, i_S$ become $\mathbf{a}' = (0, 0, 0, 0, 0, 0, 1, 3, 5)$, $\mathbf{b}' = (0, 0, 0, 0, 0, 0, -4, -5, -3)$ and $i'_S = 7$. Assume instead that a new buy limit order arrives one tick away from the best ask price, then $\mathbf{a}' = (1, 3, 5, 4, 2, 4, 4, 4, 4)$, $\mathbf{b}' = (-1, 0, 0, 0, -1, 0, -4, -5, -3)$ and $i'_S = 1$. Extracted from Abergel and Jedidi (2013)

## Arrival of Orders

The state of the order book evolves under the action of the agents operating on the market *via* the following events:

- arrival of a new market order;
- arrival of a new limit order;
- cancellation of an already existing limit order.

These events are described by *independent* Poisson processes:

- $M^{\pm}(t)$: Counting processes of market orders, with constant intensities $\lambda^{M^+}$ and $\lambda^{M^-}$;
- $L_i^{\pm}(t)$: Counting processes of limit orders at level $i$, with constant intensities $\lambda_i^{L^{\pm}}$;
- $C_i^{\pm}(t)$: Counting processes of cancellations of limit orders at level $i$, with stochastic intensities $\lambda_i^{C^+} a_i$ and $\lambda_i^{C^-}|b_i|$.

The superscript "+" (respectively "−") refers to the ask (respectively bid) side of the book. Also, as already mentioned in Paragraph 6.2.1, all orders have a fixed unit size $q$. This assumption is convenient to carry out our analysis and is, for now, of secondary importance in the general questions we are addressing. Its influence on the limit order book - particularly, its shape - will be put under careful scrutiny in Chapter 7.

Note that the intensity of the cancellation process at level $i$ is *proportional* to the available quantity at that level. This assumption is equivalent to saying that each order at level $i$ has a lifetime drawn from an exponential distribution with intensity $\lambda_i^{C^{\pm}}$. This strict proportionality assumption will be somewhat relaxed in Section 6.4. Note finally that buy limit orders $L_i^-(t)$ arrive below the ask price $P^A(t)$, and sell limit orders $L_i^+(t)$ arrive above the bid price $P^B(t)$.

### 6.2.2 Evolution of the order book

We can write the following coupled stochastic differential equations (SDE) for the quantities of outstanding limit orders on each side of the order book:

$$\begin{aligned} da_i(t) &= -\mathbf{1}_{\{a_i(t)\neq 0\}} (q - \mathbf{A}_{i-1})_+ dM^+(t) + qdL_i^+(t) - qdC_i^+(t) \\ &+ \left(J^{M^-}(\mathbf{a}) - \mathbf{a}\right)_i dM^-(t) + \sum_{i=1}^{K} \left(J^{L_i^-}(\mathbf{a}) - \mathbf{a}\right)_i dL_i^-(t) \\ &+ \sum_{i=1}^{K} \left(J^{C_i^-}(\mathbf{a}) - \mathbf{a}\right)_i dC_i^-(t), \end{aligned} \quad (6.19)$$

and

$$db_i(t) = \mathbf{1}_{\{b_i(t) \neq 0\}} (q - \mathbf{B}_{i-1})_+ dM^-(t) - q dL_i^-(t) + q dC_i^-(t)$$
$$+ \left(J^{M^+}(\mathbf{b}) - \mathbf{b}\right)_i dM^+(t) + \sum_{i=1}^{K} \left(J^{L_i^+}(\mathbf{b}) - \mathbf{b}\right)_i dL_i^+(t)$$
$$+ \sum_{i=1}^{K} \left(J^{C_i^+}(\mathbf{b}) - \mathbf{b}\right)_i dC_i^+(t) \qquad (6.20)$$

(remember that, by convention, the $b_i$s are non-positive). In Eqs (6.19) and (6.20), the first three terms describe in a straightforward manner the evolution of the queue at a given limit $i$ under the influence of the three type of events that can directly affect it:

- A buy market order decreasing by an amount $q$ the first non-zero limit on the ask side, possibly hitting the liquidity reservoir if all visible limits are empty (and similarly on the bid side);
- A new limit order increasing by an amount $q$ the corresponding limit;
- A cancellation order decreasing by an amount $q$ the corresponding limit.

By assumption, the intensity of the point processes triggering a cancellation is 0 when the corresponding quantity is 0, avoiding all inconsistencies. As for the market orders, no such assumption is made, hence, the use of the indicator function.

As for the $J$'s, they are *shift operators* corresponding to the renumbering of the ask side following an event affecting the bid side of the book and vice versa. For instance the shift operator corresponding to the arrival of a sell market order $dM^-(t)$ of size $q$ is

$$J^{M^-}(\mathbf{a}) = \Big(\underbrace{0, 0, \ldots, 0}_{k \text{ times}}, a_1, a_2, \ldots, a_{K-k}\Big), \qquad (6.21)$$

with

$$k = \inf\{p : \sum_{j=1}^{p} |b_j| > q\} - \inf\{p : |b_p| > 0\} \qquad (6.22)$$
$$= \mathbf{A}^{-1}(q) - i_S$$

[with the notations introduced in Eqs (6.14) and (6.15)], expressing the fact that the limit order book always has exactly $K$ visible limits, and that the reference price for the ask side of the book possibly changes if a sell market order eats up all the available liquidity at the best bid price. Similarly, the cancellation of a limit order *at the best bid* will have exactly the same effect on the ask side:

$$J^{C^-_{i_S}}(\mathbf{a}) = J^{M^-}(\mathbf{a}), \qquad (6.23)$$

whereas a cancellation at other limits on the bid side has no effect on the ask side. Finally, a new buy limit order *within the spread*, $k$ ticks away from the previous best bid price, will shift the ask side according to

$$J^{L^-_i}(\mathbf{a}) = \left(a_{1+k}, a_{2+k}, \ldots, a_K, \underbrace{a_\infty, \ldots, a_\infty}_{k \text{ times}}\right), \qquad (6.24)$$

with

$$k := i_S - i > 0.$$

Of course, similar expressions can be derived for the shift operators acting on the bid side of the order book.

In the next sections, we will study some general properties of the order book, starting with the generator associated with this $2K$-dimensional continuous-time Markov chain.

### 6.2.3 Infinitesimal generator

The following result characterizes the generator associated to the Markovian point process driving the order book (basic definitions concerning point processes are recalled in Appendix C, see also Brémaud (1981) Daley and Vere-Jones (2003) Daley and Vere-Jones (2008) for an in-depth treatment).

**Proposition 6.1** *The infinitesimal generator associated to the dynamics of the limit order book is the operator $\mathcal{L}$ defined by*

$$\mathcal{L}f(\mathbf{a};\mathbf{b}) = \lambda^{M^+}\left(f\left([a_i - (q - A(i-1))_+]_+; J^{M^+}(\mathbf{b})\right) - f\right)$$

$$+ \sum_{i=1}^{K} \lambda^{L^+}_i \left(f\left(a_i + q; J^{L^+_i}(\mathbf{b})\right) - f\right)$$

$$+ \sum_{i=1}^{K} \lambda^{C^+}_i a_i \left(f\left(a_i - q; J^{C^+_i}(\mathbf{b})\right) - f\right)$$

$$+ \lambda^{M^-}\left(f\left(J^{M^-}(\mathbf{a}); [b_i + (q - B(i-1))_+]_-\right) - f\right)$$

$$+ \sum_{i=1}^{K} \lambda_i^{L^-} \left( f\left( J^{L_i^-}(\mathbf{a}); b_i - q \right) - f \right)$$

$$+ \sum_{i=1}^{K} \lambda_i^{C^-} |b_i| \left( f\left( J^{C_i^-}(\mathbf{a}); b_i + q \right) - f \right), \qquad (6.25)$$

where we write $f(a_i; \mathbf{b})$ instead of $f(a_1, \ldots, a_i, \ldots, a_K; \mathbf{b})$ etc. to ease the notations, and

$$x_+ := \max(x, 0), \qquad x_- := \min(x, 0), x \in \mathbb{R}. \qquad (6.26)$$

The operator above, although cumbersome to put in writing, is simple to decipher: A series of standard difference operators corresponding to the "deposition-evaporation" of orders at each limit, combined with the shift operators expressing the moves in the best limits and therefore, in the origins of the frames for the two sides of the order book. Note the coupling of the two sides: the shifts on the $a$s depend on the $b$s, and vice versa. More precisely the shifts depend on the profile of the order book on the other side. Also, one can easily check that the formulation with an indicator function for the case of empty limits is equivalent to its reformulation with the nested positive parts.

### 6.2.4 Price dynamics

We now focus on the dynamics of the best ask and bid prices, denoted by $P^A(t)$ and $P^B(t)$. One can write the following SDE:

$$\begin{aligned} dP^A(t) &= \Delta P[\left( \mathbf{A}^{-1}(q) - i_S \right) dM^+(t) \\ &\quad - \sum_{i=1}^{K} (i_S - i)_+ dL_i^+(t) + \left( \mathbf{A}^{-1}(q) - i_S \right) dC_{i_S}^+(t)] \end{aligned} \qquad (6.27)$$

and

$$\begin{aligned} dP^B(t) &= -\Delta P[\left( \mathbf{B}^{-1}(q) - i_S \right) dM^-(t) \\ &\quad - \sum_{i=1}^{K} (i_S - i)_+ dL_i^-(t) + \left( \mathbf{B}^{-1}(q) - i_S \right) dC_{i_S}^-(t)], \end{aligned} \qquad (6.28)$$

which describe the various events that affect them: change due to a market order, change due to limit orders inside the spread, and change due to the cancellation of a limit order at the best price. Equivalently, the respective dynamics of the mid-price and the spread are:

$$dP(t) = \frac{\Delta P}{2}\left[\left(\mathbf{A}^{-1}(q) - i_S\right)dM^+(t) - \left(\mathbf{B}^{-1}(q) - i_S\right)dM^-(t)\right.$$
$$- \sum_{i=1}^{K}(i_S - i)_+ dL_i^+(t) + \sum_{i=1}^{K}(i_S - i)_+ dL_i^-(t)$$
$$\left. + \left(\mathbf{A}^{-1}(q) - i_S\right)dC_{i_S}^+(t) - \left(\mathbf{B}^{-1}(q) - i_S\right)dC_{i_S}^-(t)\right], \tag{6.29}$$

$$dS(t) = \Delta P\left[\left(\mathbf{A}^{-1}(q) - i_S\right)dM^+(t) + \left(\mathbf{B}^{-1}(q) - i_S\right)dM^-(t)\right.$$
$$- \sum_{i=1}^{K}(i_S - i)_+ dL_i^+(t) - \sum_{i=1}^{K}(i_S - i)_+ dL_i^-(t)$$
$$\left. + \left(\mathbf{A}^{-1}(q) - i_S\right)dC_{i_S}^+(t) + \left(\mathbf{B}^{-1}(q) - i_S\right)dC_{i_S}^-(t)\right]. \tag{6.30}$$

The equations above are interesting in that they relate in an explicit way the profile of the order book to the size of an increment of the mid-price or the spread, therefore linking the price dynamics to the order flow. For instance the conditional infinitesimal drifts of the mid-price and the spread, given the shape of the order book at time $t$, are given by:

$$\mathbb{E}\left[dP(t)|\,(\mathbf{a};\mathbf{b})\right] = \frac{\Delta P}{2}\left[\left(\mathbf{A}^{-1}(q) - i_S\right)\lambda^{M^+} - \left(\mathbf{B}^{-1}(q) - i_S\right)\lambda^{M^-}\right.$$
$$- \sum_{i=1}^{K}(i_S - i)_+ \lambda_i^{L^+} + \sum_{i=1}^{K}(i_S - i)_+ \lambda_i^{L^-}$$
$$\left. + \left(\mathbf{A}^{-1}(q) - i_S\right)\lambda_{i_S}^{C^+} a_{i_S} - \left(\mathbf{B}^{-1}(q) - i_S\right)\lambda_{i_S}^{C^-}|b_{i_S}|\right]dt, \tag{6.31}$$

$$\mathbb{E}\left[dS(t)|\,(\mathbf{a};\mathbf{b})\right] = \Delta P\left[\left(\mathbf{A}^{-1}(q) - i_S\right)\lambda^{M^+} + \left(\mathbf{B}^{-1}(q) - i_S\right)\lambda^{M^-}\right.$$
$$- \sum_{i=1}^{K}(i_S - i)_+ \lambda_i^{L^+} - \sum_{i=1}^{K}(i_S - i)_+ \lambda_i^{L^-}$$
$$\left. + \left(\mathbf{A}^{-1}(q) - i_S\right)\lambda_{i_S}^{C^+} a_{i_S} + \left(\mathbf{B}^{-1}(q) - i_S\right)\lambda_{i_S}^{C^-}|b_{i_S}|\right]dt. \tag{6.32}$$

## 6.3 Ergodicity and Diffusive Limit

In this section, our interest lies in the following questions:

(i) Is the order book model defined above *stable*?

(ii) What is the *stochastic-process limit* of the price at large time scales?

### 6.3.1 Ergodicity of the order book

Denote by $Q^t(\mathbf{X},.)$ the transition probability function at time $t$ of the Markov process $X_t$ starting from $X$ at time 0, and by $\|\mu\|$ the total variation norm of a probability measure $\mu$ (see Appendix C.2.1 for details). Then, the following result holds:

**Theorem 6.2** *If $\underline{\lambda_C} = \min_{1 \leq i \leq K}\{\lambda_i^{C\pm}\} > 0$, then $(\mathbf{X}(t))_{t \geq 0} = (\mathbf{a}(t); \mathbf{b}(t))_{t \geq 0}$ is an ergodic Markov process. In particular $(\mathbf{X}(t))$ has a unique stationary distribution $\Pi$. Moreover, the rate of convergence of the order book to its stationary state is exponential. That is, there exist $r$, $0 < r < 1$, and $R < \infty$ such that*

$$\|Q^t(\mathbf{X},.) - \Pi(.)\| \leq Rr^t V(\mathbf{X}), t \in \mathbb{R}^+, \mathbf{X} \in \mathcal{S}. \tag{6.33}$$

**Proof** Let

$$V(\mathbf{X}) := V(\mathbf{a};\mathbf{b}) := \sum_{i=1}^K a_i + \sum_{i=1}^K |b_i| + q \tag{6.34}$$

be the total number of shares in the book ($+q$ shares). $V$ is a positive function and tends to $+\infty$ as $\mathbf{a};\mathbf{b}$ tend to $\infty$: in other words, $V$ is coercive.

Using the expression of the infinitesimal generator Eq. (6.25) we have

$$\mathcal{L}V(\mathbf{X}) \leq -\left(\lambda^{M^+} + \lambda^{M^-}\right)q + \sum_{i=1}^K \left(\lambda_i^{L^+} + \lambda_i^{L^-}\right)q - \sum_{i=1}^K \left(\lambda_i^{C^+} a_i + \lambda_i^{C^-}|b_i|\right)q$$

$$+ \sum_{i=1}^K \lambda_i^{L^-}(i_S - i)_+ a_\infty + \sum_{i=1}^K \lambda_i^{L^+}(i_S - i)_+ |b_\infty| \tag{6.35}$$

$$\leq -\left(\lambda^{M^+} + \lambda^{M^-}\right)q + \left(\Lambda^{L^-} + \Lambda^{L^+}\right)q - \underline{\lambda^C} qV(\mathbf{X})$$

$$+ K\left(\Lambda^{L^-} a_\infty + \Lambda^{L^+}|b_\infty|\right), \tag{6.36}$$

where

$$\Lambda^{L^\pm} := \sum_{i=1}^K \lambda_i^{L^\pm} \text{ and } \underline{\lambda^C} := \min_{1 \leq i \leq K}\{\lambda_i^{C\pm}\} > 0. \tag{6.37}$$

The first three terms on the right hand side of inequality Eq. (6.35) correspond respectively to the arrival of a market, limit or cancellation order, ignoring the effect of the shift operators. The last two terms are due to shifts occurring after the arrival of a limit order inside the spread. The terms due to shifts occurring after market or cancellation orders (which we do not put in the RHS of Eq. (6.35)) are negative, hence, the inequality. To obtain inequality Eq. (6.36), we used the fact that the spread $i_S$ is bounded by $K + 1$ - a consequence of the boundary conditions we impose - and hence $(i_S - i)_+$ is bounded by $K$.

The drift condition Eq. (6.36) can be rewritten as

$$\mathcal{L}V(\mathbf{X}) \leq -\beta V(\mathbf{X}) + \gamma, \tag{6.38}$$

which is readily improved to

$$\mathcal{L}V(\mathbf{X}) \leq -\beta V(\mathbf{X}) + \gamma \mathbf{1}_\mathbf{K} \tag{6.39}$$

for some positive constants $\beta$, $\gamma$ and a compact subset $\mathbf{K}$ of the state space, thanks to the coercivity of the Lyapunov function $V$. Inequality Eq. (6.39), together with the countability of the state space ensuring that all compact sets are *petite* sets in the sense of Meyn and Tweedie (2009) (see Appendix C.2.1 for a discussion and precise references) for a discussion and precise references) let us assert that $(\mathbf{X})$ is $V$-uniformly ergodic, hence Eq. (6.33).

As an easy consequence, there holds the following corollary:

**Corollary 6.3** *The spread $S(t) = i_S \Delta P$ has a well-defined stationary distribution.*

## 6.3.2 Large-scale limit of the price process

This section is devoted to the asymptotics of the suitably rescaled, centered price process.

Such limit theorems are actively researched of late: In Cont and de Larrard (2012), the authors show that in a simplified model of an order book for a liquid stock, one can derive a diffusive limit for the volume at the best quotes of the book. These types of results are also the focus of Guo et al. (2015), which also deals with the relative position of an order inside the queue. Other limit theorems are obtained in Horst and Paulsen (2015). The authors prove that under some simplifying assumptions and a specific scaling, one may obtain in the limit coupled differential equations for the bid and ask price dynamics, as well as the for the shape of the order book. A somewhat related description of the limiting shapes of the order book as solution of differential equations is also the subject of Gao et al. (2014).

Our treatment of these questions combines the ergodic theory of Markov processes with martingale convergence theorems. This approach is extremely general and flexible, and prone to many generalizations for Markovian models of limit order books. We state and prove our main result concerning the long-time price dynamics in the case of Poissonian arrival times. The case of more general drivers will be studied in Chapter 8.

For notational convenience, we recast the equation for the price, or rather, any of the prices, under the general form

$$P_t = \int_0^t \sum_i F_i(\mathbf{X}(u)) \, dN^i(u), \qquad (6.40)$$

where the $N^i$'s are the point processes driving the events affecting the limit order book, $v^i \equiv v^i(\mathbf{X})$ is the (possibly state-dependent) intensity of $N^i$, and the $F_i$'s are the jumps in the price of interest when process $N^i$ jumps.

Denote by $\Pi$ the stationary distribution of $\mathbf{X}$ as provided by Proposition 6.2. Using the Ergodic Theorem C.7 together with the Martingale Convergence Theorem C.8, one can show the following proposition:

**Proposition 6.4** *Consider the price process described by Eq. (6.40) above, and introduce the sequence of martingales $\hat{P}^n$ formed by the centered, rescaled price*

$$\hat{P}^n(t) \equiv \frac{P(nt) - Q(nt)}{\sqrt{n}},$$

*where $Q$ is the predictable compensator of $P$*

$$Q_t = \sum_i \int_0^t v^i(\mathbf{X}(s)) F_i(\mathbf{X}(s)) \, ds.$$

*Then, $\hat{P}^n$ converges in distribution to a Wiener process $\hat{\sigma} W$, where the volatility $\hat{\sigma}$ is given by*

$$\hat{\sigma}^2 = \lim_{t \to +\infty} \frac{1}{t} \sum_i \int_0^t v^i(\mathbf{X}(s)) (F_i(\mathbf{X}(s)))^2 \, ds$$

$$= \sum_i \int v^i(\mathbf{X}) (F_i(\mathbf{X}))^2 \, \Pi(d\mathbf{X}) \qquad (6.41)$$

*(where we use, with a slight abuse of notations, the same letter for a process and the corresponding state variable in the state space).*

**Proof** Proposition 6.4 will follow from the convergence of the predictable quadratic variation of $\hat{P}^n$. By construction, there holds

$$<\hat{P}^n, \hat{P}^n>(t) = \frac{1}{n} \sum_i \int_0^{nt} v^i(\mathbf{X}(s)) (F_i(\mathbf{X}(s)))^2 \, ds,$$

or else
$$<\hat{P}^n, \hat{P}^n>(t) = t\Big(\frac{1}{n}\sum_i \int_0^{nt} v^i(\mathbf{X}(s))(F_i(\mathbf{X}(s)))^2\,ds\Big),$$

and Theorem C.7 ensures that

$$\lim_{t\to+\infty}^{a.s.} \frac{1}{nt}\sum_i \int_0^{nt} v^i(\mathbf{X}(s))(F_i(\mathbf{X}(s)))^2\,ds$$

$$= \sum_i \int v^i(\mathbf{X})(F_i(\mathbf{X}))^2\,\Pi(d\mathbf{X})$$

whenever the integrability conditions in Theorem C.7 are satisfied. Now, those are easily seen to hold true, since, the integrand in the predictable quadratic variation is a bounded function. As a matter of fact, the only possibly unbounded term would come from the intensity of cancellation orders, proportional to the $a_i, |b_i|$'s. However, whenever a cancellation order causes a price change, then necessarily, the book is in a state where the quantity at the best limit that moves is precisely equal to $q$. Hence, the boundedness follows.

The other condition for the martingale convergence theorem to apply is trivially satisfied, since, the size of the jumps of $\hat{P}^n$ is bounded by $\frac{C}{\sqrt{n}}$, $C$ being some constant.

Appealing as it first seems, Proposition 6.4 is not satisfactory: In order to give a more precise characterization of the dynamics of the rescaled price process, it is necessary to understand thoroughly the behaviour of its compensator $Q_{nt}$. As a matter of fact, $Q_{nt}$ itself satisfies an ergodic theorem, and if its asymptotic variance is not negligible with respect to $nt$, one cannot conclude directly from Proposition 6.4 that the rescaled price process $\frac{P_{nt}}{\sqrt{n}}$ behaves like a Wiener process with a deterministic drift.

The next result provides a more accurate answer, valid under general ergodicity conditions.

**Theorem 6.5** *Write as above the price*

$$P_t = \sum_i \int_0^t F_i(\mathbf{X}(s))\,dN^i(s)$$

*and its compensator*

$$Q_t = \sum_i \int_0^t v^i(\mathbf{X}(s))\,F_i(\mathbf{X}(s))\,ds.$$

*Define*

$$h = \sum_i v^i(\mathbf{X}) F_i(\mathbf{X})$$

*and let*

$$\alpha \stackrel{a.s.}{=} \lim_{t\to+\infty} \frac{1}{t} \sum_i \int_0^t v^i(\mathbf{X}(s))(F_i(\mathbf{X}(s)))\,ds = \int h(\mathbf{X})\Pi(d\mathbf{X}).$$

*Finally, introduce the solution g to the Poisson equation*

$$\mathcal{L}g = h - \alpha \tag{6.42}$$

*and the associated martingale*

$$Z_t = g(\mathbf{X}(t)) - g(\mathbf{X}(0)) - \int_0^t \mathcal{L}g(\mathbf{X}(s))ds \equiv g(\mathbf{X}(t)) - g(\mathbf{X}(0)) - Q(t) + \alpha t.$$

*Then, the deterministically centred, rescaled price*

$$\bar{P}^n(t) \equiv \frac{P(nt) - \alpha nt}{\sqrt{n}}$$

*converges in distribution to a Wiener process $\bar{\sigma}W$. The asymptotic volatility $\bar{\sigma}$ satisfies the identity*

$$\bar{\sigma}^2 = \lim_{t\to+\infty} \frac{1}{t} \sum_i \int_0^t v^i(\mathbf{X}(s)) \left((F_i - \Delta^i(g))(\mathbf{X}(s))\right)^2 ds \tag{6.43}$$

$$\equiv \sum_i \int v^i(\mathbf{X}) \left((F_i - \Delta^i(g))(\mathbf{X})\right)^2 \Pi(d\mathbf{X}) \tag{6.44}$$

*where $\Delta^i(g)(\mathbf{X})$ denotes the jump of the process $g(\mathbf{X})$ when the process $N^i$ jumps and the limit order book is in the state $\mathbf{X}$.*

**Proof** The martingale method, see e.g., Glynn and Meyn (1996) Duflo (1990) Jacod and Shyriaev (2003), consists in rewriting the price process under the form

$$P(t) = (P(t) - Q(t)) - Z(t) + g(\mathbf{X}(t)) - g(\mathbf{X}(0)) + \alpha t \tag{6.45}$$

$$\equiv (M(t) - Z(t)) + g(\mathbf{X}(t)) - g(\mathbf{X}(0)) + \alpha t,$$

so that

$$\bar{P}^n(t) = \frac{\tilde{M}(nt) + g(\mathbf{X}(t)) - g(\mathbf{X}(0))}{\sqrt{n}},$$

where $\tilde{M} = M - Z$ is a martingale. Therefore, the theorem is proven if (see Glynn and Meyn (1996), Theorem 4.2 or Bhattacharya (1982)) one can show that $\frac{g(\mathbf{X}(t)) - g(\mathbf{X}(0))}{\sqrt{n}}$ converges to 0 in $L^2(\Pi(d\mathbf{X}))$, or simply, that $g \in L^2(\Pi(d\mathbf{X}))$.

Theorem 4.4 of Glynn and Meyn (1996) states that the condition

$$h^2 \leqslant V \qquad (6.46)$$

(where $V$ is a Lyapunov function for the process) is sufficient for $g$ to be in $L^2(\Pi(d\mathbf{X}))$ - however, Condition Eq. (6.46) is trivially satisfied since, $h$ is bounded.

### 6.3.3 Interpreting the asymptotic volatility

A general formula for the low frequency volatility of the price process is provided in Eq. (6.43); it is related to the frequency of events that cause a price change, and to the size of price jumps when a change occurs. Although, Formula (6.43) can easily be implemented numerically by using its formulation as a time average, its analytical computation requires the knowledge of the stationary distribution of the order book. However, some simplifying hypotheses help shed some light on its interpretation and qualitative dependency on the model parameters. Assume for instance that one is interested in modelling *large tick* assets, for which the price change is *always* equal to 1 tick. In our framework, this is made possible by choosing $K = 1$: only one limit on each side of the order book is modelled. In this case, all the $F_i$'s introduced in Section 6.3.2 are equal to 1 or 0, and the asymptotic variance can be rewritten by separating the events that change the price from those that do not.

Similarly to the empirical approach presented in Section 4.4.1, let us classify market, limit and cancellation orders depending on whether they change the price or not, using a 1 (resp. 0) superscript to indicate that the event changes (resp. does not change) the price:

$$M^{\pm} = M^{\pm,1} + M^{\pm,0},$$

$$L_i^{\pm} = L_i^{\pm,1} + L_i^{\pm,0},$$

$$C_i^{\pm} = C_i^{\pm,1} + C_i^{\pm,0}.$$

Now, should all these processes be independent Poisson processes, the asymptotic variance would be given using Eqs (6.41) or (6.43) (see comment below) by

$$\bar{\sigma}^2 = (\Delta P)^2 \left( \lambda^{M^+,1} + \lambda^{M^-,1} + \sum_i \left( \lambda_i^{L^+,1} + \lambda_i^{L^-,1} \right) + \lambda_{is}^{C^+,1} + \lambda_{is}^{C^-,1} \right),$$

where all the quantities involved are easily interpreted, and can be measured empirically from the data.

Another interesting question concerning Formula (6.43) is the role played by the correcting term coming from the solution $g$ to the Poisson Eq. (6.42). In the case of Poisson arrival for the price-changing processes and deterministic price changes, the right-hand-side of Eq. (6.42). is 0, so that the correcting terms are also 0: Formulae (6.41) and (6.43) coincide. In general this is not the case, and one should find an estimate of the correcting terms - essentially, a control of the variance of $h = \sum_i v^i(\mathbf{X}) F_i((X))$ when the $\lambda_i$s are now random. This more general case is analytically very intricate, although easily attainable *via* numerical simulations.

## 6.4 The Role of Cancellations

In this short section, we address in more generality the role played by the cancellation rate in the ergodicity and price diffusivity of the limit order book model introduced in Section 6.2. The results presented here rely on the use of a more general Lyapunov function, and require a less stringent condition on the cancellation rate.

Assume now that in the set of assumptions listed in 6.2.1, those concerning the cancellation rates are modified as follows:

- $C_i^\pm(t)$: Cancellation of a limit order at level $i$, with intensity $\lambda_i^{C^+}(\mathbf{a};\mathbf{b})$ and $\lambda_i^{C^-}(\mathbf{a};\mathbf{b})$, where the functions $\lambda_i^{C^+}$ and $\lambda_i^{C^-}$ are positive, bounded away from 0 and tend to $\infty$ as $(\mathbf{a};\mathbf{b}) \to \infty$.

Then, there holds the

**Proposition 6.6** *The properties stated in Theorem 6.2 and Theorem 6.5 hold without change.*

**Proof** The method is exactly the same, based on the use of an *adhoc* Lyapunov function. Of course, the linear function $V$ introduced in the proof of Theorem 6.2 (see (8.5)) does not work under the general assumptions 6.4. However, it is straightforward to check that the function

$$\tilde{V}(\mathbf{X}) := \exp(V(\mathbf{X})) \tag{6.47}$$

actually solves the problem: Upon calculating $\mathcal{L}\tilde{V}$ as in the proof of Theorem 6.2, one easily sees that the exponential factors out and - since by assumption the intensity of incoming limit orders is dominated by that of the cancellation orders - that an inequality of the form

$$\mathcal{L}\tilde{V} \leqslant -\beta'\tilde{V} + \gamma' \tag{6.48}$$

(compare to Eq. (6.39)) holds for some positive constants $\beta', \gamma'$. Consequently, the ergodicity of the limit order book is proven in a similar fashion. As for the dynamics of the rescaled price process and its convergence to a Wiener process, one simply observes again that the RHS $h$ of the Poisson Eq. (6.42) is still a bounded function, so that one can safely apply Theorem 4.4 in Glynn and Meyn (1996) to obtain the FCLT exactly as in the proof of Theorem 6.5.

## 6.5 Conclusion

In this chapter, we have analysed a simple Markovian order book model, in which elementary changes in the price and spread processes are explicitly linked to the instantaneous shape of the order book and the order flow parameters. Our assumptions are: independent arrivals of orders of different types, strong intensity of cancellations, constant order sizes, and the presence of two reservoirs of liquidity $K$ ticks away from the best quotes.

Two fundamental properties were investigated: the ergodicity of the order book and the large-scale limit of the price process. The first property is desirable in that it assures the stability of the order book in the long run, and gives a theoretical underpinning to statistical measurements on order book data. The second addresses the natural question of the behaviour of the price sampled at lower frequency, and relates it to a Wiener process. In a sense, this chapter serves as a mathematical justification to the simple Bachelier model of asset prices, from a market microstructure perspective.

We believe that the approach presented here is interesting mainly for the introduction of a general framework and a set of mathematical tools well-suited to further investigations of more sophisticated models. Some results in this direction will be presented in Chapter 8. Meanwhile, the next chapter focuses on a different, yet also quite natural question: that of the shape of a limit order book, and its sensitivity to the size of incoming orders.

CHAPTER 7

# The Order Book as a Queueing System

## 7.1 Introduction

In this chapter, we move forward on the mathematical study of zero-intelligence models by deriving some analytical properties of a limit order book under the assumptions introduced in Section 6.2. The model is then extended to the case of random order sizes, thereby allowing to study the relationship between the size of the incoming limit orders and the shape of the order book.

Recall that the shape of the order book is simply the function which for any price gives the number of shares standing in the order book at that price; and the cumulative shape up to price $p$ is the total quantity offered in the order book between the best limit and price $p$. If the limit order book is ergodic, the (cumulative) shape admits a stationary distribution, and its expectation with respect to this stationary distribution will be simply called the average (cumulative) shape.

Understanding the average shape of an order book and its link to the order flows is not straightforward. The first empirical observations of the shape of an order book stated that, at least for the first limits, the shape of the order book is increasing away from the spread (Biais et al. 1995). With better data, one can complement this view and state that "the average order book has a maximum away from the current bid/ask, and a tail reflecting the statistics of the incoming orders" (Bouchaud et al. 2002; Potters and Bouchaud, 2003), i.e. that the limit order book is *hump-shaped*. The decrease of the tail of the order book is difficult to estimate because one needs complete data, including limit orders submitted far away from the best quotes, which is often not disclosed by exchanges. A power law decrease (Bouchaud et al. 2002) or an exponential decrease (Gu et al. 2008) or even a whole lognormal shape (Preis et al. 2006) have been suggested.

The hump-shaped order book appears quite easily in simulations, as we will see later in Chapter 9. Studying simulation results, Cont et al. (2010) observe that "the average profile of the order book displays a hump [...] that does not result from any fine-tuning of model parameters or additional ingredients such as correlation between order flow and past price moves", but no explicit link between the average shape and the order flows is made. The results presented in this chapter hopefully closes this gap, at least partially.

## 7.2 A Link Between the Flows of Orders and the Shape of an Order Book

### 7.2.1 The basic one-sided queueing system

The aim of this section is to present the basic one-sided order book model, discuss the relevance of its assumptions and recall some results from queueing theory in the context of this order book model. Let us consider a one-sided order book model, i.e. a model in which all limit orders are ask orders, and all market orders are buy orders. Bid price is assumed to be constantly equal to zero, and consequently spread and ask price are identically equal. From now on, this quantity will be simply referred to as the price. We will use the notations already introduced in Chapter 6, sometimes slightly simplified to the one-sided setting of this chapter. Let $P^A(t)$ denote the price at time $t$. $\{P^A(t), t \in [0 \in \infty)\}$ is a continuous-time stochastic process with value in the discrete set $\{1, \ldots, K\}$. In other words, the price is given in number of ticks. Let $\Delta P$ be the tick size, such that the price range of the model in currency is actually $\{\Delta P, \ldots, K\Delta P\}$. For realistic modelling and empirical fitting performance, one may assume that the maximum price $K$ is chosen very large, but in fact it will soon be obvious that this upper bound does not affect in any way the order book for lower prices. For all $i \in \{1, \ldots, K\}$, (ask) limit orders at price $i$ are submitted according to a Poisson process with parameter $\lambda_i^L$ (we drop the $\pm$ of the notation introduced in Chapter 6, since we only deal with one side of the book and all limit orders are ask orders). These processes as assumed to be mutually independent, so that the number of orders submitted at prices $1 \ldots, r$ is a Poisson process with parameter $\Lambda_r^L$ defined as $\Lambda_r^L = \sum_{i=1}^{r} \lambda_i^L$. All limit orders standing in the book may be cancelled. It is assumed that the time intervals between submission and cancellation form a set of mutually independent random variables identically distributed according to an exponential distribution with parameter $\lambda^C > 0$ (here again we simplify the notation of Chapter 6, dropping the unnecessary index and $\pm$ symbol). Finally, (buy) market orders are submitted at random times according to a Poisson process with parameter $\lambda^M$. Note that all orders are assumed to be of unit size. This restriction will be dropped in Sections 7.4 *et sq.*

Still using the notations of Chapter 6, let $\{\mathbf{A}_k(t), t \in [0, \infty)\}$ be the stochastic process representing the number of limit orders at prices $1, \ldots, k$ standing in the order book at time $t$. $\mathbf{A}_k$ is thus the cumulative shape of the order book in our model. $\mathbf{A}_k$ can be viewed as a birth-death process with birth rate $\Lambda_k^L$ and death rate $\lambda^M + n\lambda^C$ in state $n$; it may equivalently be viewed as the size of a $M/M/1+M$ queueing system with arrival rate $\Lambda_k^L$, service rate $\lambda^M$ and reneging rate $\lambda^C$ (see e.g., (Feller, 1968, Chapter XVII) or (Brémaud, 1999, Chapter 8) among many textbook references). This queueing system will now be refered to as the $1 \to k$ queueing system. $\mathbf{A}_k$ admits a stationary distribution $\pi_{\mathbf{A}_k}(\cdot)$ as soon as $\lambda^C > 0$. The matrix form of the infinitesimal generator is written:

$$\begin{pmatrix} -\Lambda_k^L & \Lambda_k^L & 0 & 0 & 0 & \cdots \\ \lambda^M + \lambda^C & -(\Lambda_k^L + \lambda^M + \lambda^C) & \Lambda_k^L & 0 & 0 & \cdots \\ 0 & \lambda^M + 2\lambda^C & -(\Lambda_k^L + \lambda^M + 2\lambda^C) & \Lambda_k^L & 0 & \cdots \\ \vdots & \vdots & & \ddots & \ddots & \ddots \end{pmatrix}. \quad (7.1)$$

The infinitesimal generator is here conveniently written in matrix form in this discrete setting, but note that it is equivalent to the functional operator form $\mathcal{L}$ used in Chapter 6. The stationary probability $\pi_{\mathbf{A}_k}$ is classically obtained and written for all $n \in \mathbb{N}^*$:

$$\pi_{\mathbf{A}_k}(n) = \pi_{\mathbf{A}_k}(0) \prod_{i=1}^{n} \frac{\Lambda_k^L}{\lambda^M + i\lambda^C}, \quad (7.2)$$

and setting $\sum_{n=0}^{\infty} \pi_{\mathbf{A}_k}(n) = 1$ gives:

$$\pi_{\mathbf{A}_k}(0) = \left( \sum_{n=0}^{\infty} \prod_{i=1}^{n} \frac{\Lambda_k^L}{\lambda^M + i\lambda^C} \right)^{-1}. \quad (7.3)$$

Introducing the normalized parameters $\bar{\Lambda}_k^L = \frac{\Lambda_k^L}{\lambda^C}$ and $\bar{\lambda}^M = \frac{\lambda^M}{\lambda^C}$, and after some simplifications, we write for all $n \in \mathbb{N}$:

$$\pi_{\mathbf{A}_k}(n) = \frac{e^{-\bar{\Lambda}_k^L} \left( \bar{\Lambda}_k^L \right)^{\bar{\lambda}^M}}{\bar{\lambda}^M \Gamma_{\bar{\Lambda}_k^L}(\bar{\lambda}^M)} \prod_{i=1}^{n} \frac{\bar{\Lambda}_k^L}{i + \bar{\lambda}^M}, \quad (7.4)$$

where $\Gamma_y$ is the lower incomplete version of the Euler gamma function:

$$\Gamma_y : \mathbb{R}_+ \to \mathbb{R}, x \mapsto \int_0^y t^{x-1} e^{-t} dt. \quad (7.5)$$

Now, the price in the one-sided order book model is equal to $k$ if and only if the "$1 \to k-1$" queueing system is empty and the "$1 \to k$" system is not. Therefore, if $\mathbf{A}_k$ is distributed according to the invariant distribution $\pi_{\mathbf{A}_k}$, then the distribution $\pi_{P^A}$ of the price $P^A$ is written:

$$\pi_{p_A}(1) = 1 - \frac{e^{-\bar{\Lambda}_1^L}\left(\bar{\Lambda}_1^L\right)^{\bar{\lambda}^M}}{\bar{\lambda}^M \Gamma_{\bar{\Lambda}_1^L}(\bar{\lambda}^M)}, \quad \pi_{p_A}(K) = \frac{e^{-\bar{\Lambda}_{K-1}^L}\left(\bar{\Lambda}_{K-1}^L\right)^{\bar{\lambda}^M}}{\bar{\lambda}^M \Gamma_{\bar{\Lambda}_{K-1}^L}(\bar{\lambda}^M)}, \qquad (7.6)$$

and for all $k \in \{2, \ldots, K-1\}$,

$$\pi_{p_A}(k) = \frac{e^{-\bar{\Lambda}_{k-1}^L}\left(\bar{\Lambda}_{k-1}^L\right)^{\bar{\lambda}^M}}{\bar{\lambda}^M \Gamma_{\bar{\Lambda}_{k-1}^L}(\bar{\lambda}^M)} - \frac{e^{-\bar{\Lambda}_k^L}\left(\bar{\Lambda}_k^L\right)^{\bar{\lambda}^M}}{\bar{\lambda}^M \Gamma_{\bar{\Lambda}_k^L}(\bar{\lambda}^M)}. \qquad (7.7)$$

Using previous results, the average size $\mathbf{E}[\mathbf{A}_k]$ of the "$1 \to k$" queueing system is easily computed. From Eq. (7.4), we can write after some simplifications:

$$\mathbf{E}[\mathbf{A}_k] = \bar{\Lambda}_k^L - \frac{\Gamma_{\bar{\Lambda}_k^L}(1+\bar{\lambda}^M)}{\Gamma_{\bar{\Lambda}_k^L}(\bar{\lambda}^M)}. \qquad (7.8)$$

Still using the notations of Chapter 6, $a_k = \mathbf{A}_k - \mathbf{A}_{k-1}$ is the number of orders in the book at price $k \in \{1, \ldots, K\}$. Thus, the average shape of the order book at price $k$ is obviously:

$$\mathbf{E}[a_k] = \bar{\lambda}_k^L - \left(\frac{\Gamma_{\bar{\Lambda}_k^L}(1+\bar{\lambda}^M)}{\Gamma_{\bar{\Lambda}_k^L}(\bar{\lambda}^M)} - \frac{\Gamma_{\bar{\Lambda}_{k-1}^L}(1+\bar{\lambda}^M)}{\Gamma_{\bar{\Lambda}_{k-1}^L}(\bar{\lambda}^M)}\right), \qquad (7.9)$$

with $\bar{\lambda}_k^L = \frac{\lambda_k^L}{\lambda^C}$.

### 7.2.2 A continuous extension of the basic model

In order to facilitate the comparison with existing results, we propose a continuous version of the previous toy model. Price is now assumed to be a positive real number. Mechanisms for market orders and cancellations are identical: unit-size market orders are submitted according to a Poisson process with rate $\lambda^M$, and standing limit orders are cancelled after some exponential random time with parameter $\lambda^C$. As for the submission of limit orders, the mechanism is now slightly modified: Since, the price is continuous, instead of a finite set of homogeneous Poisson processes indexed by the number of ticks $k \in \{1, \ldots, K\}$, we now consider a spatial Poisson process on the positive quadrant $\mathbb{R}_+^2$. Let $\lambda^L(p,t)$ be a non-negative function denoting the intensity of the spatial Poisson process modelling the arrival of limit orders, the first coordinate representing the price, the second one the time (see e.g., Privault, 2013, Chapter 12 for a textbook introduction on the construction of spatial Poisson processes). As in the discrete case, this process is assumed to be time-homogeneous, and it is hence assumed that price and time are separable. Let $h_{\lambda^L} : \mathbb{R}_+ \to \mathbb{R}_+$ denote the spatial intensity function of the random events, i.e. limit orders. Then, $\lambda^L(p,t) = \alpha h_{\lambda^L}(p)$ is the intensity of the spatial Poisson process representing the arrival of limit orders.

We recall that in this framework, for any $p_1 < p_2 \in [0, \infty)$, the number of limit orders submitted at a price $p \in [p_1, p_2]$ is a homogeneous Poisson process with intensity $\int_{p_1}^{p_2} \lambda^L(p, t)\, dp$. Furthermore, if $p_1 < p_2 < p_3 < p_4$ on the real positive half-line, then the number of limit orders submitted in $[p_1, p_2]$ and $[p_3, p_4]$ form two independent Poisson processes.

Now, let $\mathbf{A}([0, p])$ be the random variable describing the cumulative size of our new order book up to price $p \in \mathbb{R}_+$. Given the preceding remarks, $\mathbf{A}([0, p])$ is, as in the previous section, the size of a $M/M/1 + M$ queueing system with arrival rate $\alpha \int_0^p h_{\lambda^L}(u)\, du$, service rate $\lambda^M$ and reneging rate $\lambda^C$. Using the results of Section 7.2.1, we obtain from Eq. (7.8):

$$\mathbf{E}[\mathbf{A}([0, p])] = \int_0^p \bar{\lambda}^L(u)\, du - f\left(\bar{\lambda}^M, \int_0^p \bar{\lambda}^L(u)\, du\right), \tag{7.10}$$

where we have defined $\bar{\lambda}^L(u) = \frac{\alpha h_{\lambda^L}(u)}{\lambda^C}$ and:

$$f(x, y) = \frac{\Gamma_y(1 + x)}{\Gamma_y(x)}. \tag{7.11}$$

From now on, $\bar{\Lambda}^L(p) = \int_0^p \bar{\lambda}^L(u)\, du$ will be the (normalized) arrival rate of limit orders up to price $p$, and $A(p) = \mathbf{E}[\mathbf{A}([0, p])]$ will be the average cumulative shape of the order book up to price $p$. Then, $a(p) = \frac{dA(p)}{dp}$ will be the average shape of the order book (per price unit, not cumulative). Straightforward differentiation of Eq. (7.10) and some terms rearrangements lead to the following proposition.

**Proposition 7.1** *In a continuous order book with homogeneous Poisson arrival of market orders with intensity $\lambda^M$, spatial Poisson arrival of limit orders with intensity $\alpha h_{\lambda^L}(p)$, and exponentially distributed lifetimes of non-executed limit orders with parameter $\lambda^C$, the average shape of the order book a is computed for all $p \in [0, \infty)$ by:*

$$a(p) = \bar{\lambda}^L(p)\left[1 - \bar{\lambda}^M \left(g_{\bar{\lambda}^M} \circ \bar{\Lambda}^L\right)(p)\left[1 - \bar{\lambda}^M[\bar{\Lambda}^L(p)]^{-1}\left[1 - \left(g_{\bar{\lambda}^M} \circ \bar{\Lambda}^L\right)(p)\right]\right]\right], \tag{7.12}$$

*where*

$$g_{\bar{\lambda}^M}(y) = \frac{e^{-y} y^{\bar{\lambda}^M}}{\bar{\lambda}^M \Gamma_y(\bar{\lambda}^M)}. \tag{7.13}$$

Let us give a few comments on the average shape we obtain. Firstly, note that by identification to Eq. (7.4), observing that $\pi_{\mathbf{A}_k}(0) = g_{\bar{\lambda}^M}(\bar{\Lambda}_k^L)$ in the discrete model, $g_{\bar{\lambda}^M}(\bar{\Lambda}^L(p))$ is to be interpreted as the probability that the order book is empty up to a price $p$. Secondly, note that letting $\bar{\lambda}^M \to 0$ in Eq. (7.12) gives $a(p) \to \bar{\lambda}^L(p)$ (cf. $\lim_{\bar{\lambda}^M \to 0} g_{\bar{\lambda}^M}(y) = e^{-y}$). Indeed, if there were no market orders, then the average shape of the order book would be equal to the normalized arrival rates. Thirdly, as $p \to \infty$, we have $a(p) \sim k\bar{\lambda}^L(p)$ for some constant $k$. This leads to our main comment, which we state as the following proposition.

> **Proposition 7.2** *The shape of the order book $a(p)$ can be written as:*
>
> $$a(p) = \bar{\lambda}^L(p)C(p), \tag{7.14}$$
>
> *where $C(p)$ is the probability that a limit order submitted at price $p$ will be cancelled before being executed.*

This proposition translates a *law of conservation of the flows of orders*: The shape of the order book is exactly the fraction of arriving limit orders that will be cancelled. The difference between the flows of arriving limit orders and the order book is exactly the fraction of arriving limit orders that will be executed.

The proof is straightforward. Indeed, in the $1 \to k$ queueing system, the average number of limit orders at price $k$ that are cancelled per unit time is $\lambda^C \mathbf{E}[a_k]$ ($\lambda^C \mathbf{E}[\mathbf{A}_k]$ is the reneging rate of $1 \to k$ queue using queueing system vocabulary). Therefore, the fraction of cancelled orders at price $k$ over arriving limit orders at price $k$, per unit time, is $C_k = \frac{\lambda^C \mathbf{E}[a_k]}{\lambda_k^L}$. Using Eq. (7.9) and some straightforward computations, the fraction $C_k$ of limit orders submitted at price $k$ which are cancelled is:

$$C_k = 1 - \frac{\bar{\lambda}^M}{\bar{\Lambda}_k^L} \left( g_{\bar{\lambda}^M}(\bar{\Lambda}_{k-1}^L) - g_{\bar{\lambda}^M}(\bar{\Lambda}_k^L) \right). \tag{7.15}$$

Therefore, in the continuous model up to price $p \in \mathbb{R}_+$ with average cumulative shape $A(p)$, the fraction of limit orders submitted at a price in $[p, p+\epsilon]$ which are cancelled is written:

$$1 - \frac{\bar{\lambda}^M}{\bar{\Lambda}^L(p+\epsilon) - \bar{\Lambda}^L(p)} \left( g_{\bar{\lambda}^M}(\bar{\Lambda}^L(p)) - g_{\bar{\lambda}^M}(\bar{\Lambda}^L(p+\epsilon)) \right). \tag{7.16}$$

By letting $\epsilon \to 0$, we obtain that the fraction $C(p)$ of limit orders submitted at price $p \in \mathbb{R}_+$ which are cancelled is:

$$C(p) = 1 - \bar{\lambda}^M g'_{\bar{\lambda}M}(\bar{\Lambda}^L(p)) \qquad (7.17)$$

$$= 1 - \bar{\lambda}^M (g_{\bar{\lambda}M} \circ \bar{\Lambda}^L)(p)\left(1 - \frac{\bar{\lambda}^M}{\bar{\Lambda}^L(p)}(1 - (g_{\bar{\lambda}M} \circ \bar{\Lambda}^L)(p))\right), \qquad (7.18)$$

which gives the final result.

This law of conservations of the flows of orders explains the relationship between the shape of the order book and the flows of arrival of limit orders. For high prices, two cases are to be distinguished. On the one hand, if the total arrival rate of limit orders is a finite positive constant $\alpha$ (for example when $h_{\lambda^L}$ is a probability density function on $[0, +\infty)$, in which case $\lim_{p \to +\infty} \bar{\Lambda}^L(p) = \int_0^\infty \lambda(u, t)\, du = \alpha \in \mathbb{R}_+^*$), then the proportionality constant between the shape of the order book $a(p)$ and the normalized limit order flow $\bar{\lambda}^L(p)$ is, as $p \to +\infty$, $C_\infty$ defined as:

$$C_\infty = \lim_{p \to \infty} C(p) = 1 - \bar{\lambda}^M g_{\bar{\lambda}M}(\alpha)\left(1 - \frac{\bar{\lambda}^M}{\alpha}(1 - g_{\bar{\lambda}M}(\alpha))\right) < 1. \qquad (7.19)$$

In such a case, the shape of the order book as $p \to +\infty$ is proportional to the normalized rate of arrival of limit orders $\bar{\lambda}^L(p)$, but not equivalent. The fraction of cancelled orders does not tend to 1 as $p \to +\infty$, i.e. market orders play a role even at high prices. On the other hand, in the case where $\lim_{p \to +\infty} \bar{\Lambda}^L(p) = \infty$, then very high prices are not reached by market orders, and the tail of the order book behaves exactly as if there were no market orders: $a(p) \sim \bar{\lambda}^L(p)$ as $p \to +\infty$. We may remark here that there exists an empirical model for the probability of execution $F(p) = 1 - C(p)$ in Mike and Farmer (2008). In this latter model, it is assumed to be the complementary cumulative distribution function of a Student distribution with parameter $s = 1.3$. As such, it is decreasing towards 0 as $p^{-s}$. In our model however, it is exponentially decreasing, and, in view of the previous discussion, does not necessarily tends towards 0.

## 7.3 Comparison to Existing Results on the Shape of the Order Book

The model presented in Section 7.2 belongs to the class of "zero-intelligence" Markovian order book models: All order flows are independent Poisson processes. Although very simple, it turns out it replicates the shapes of the order book usually obtained in previous empirical and numerical studies, as we will now see.

### 7.3.1 Numerically simulated shape in Smith et al. (2003)

A first result on the shape of the order book is provided in Smith et al. (2003), on Figs 3(a) and 3(b). These figures are obtained by numerical simulations of an order book model very similar to the one presented in Section 7.2, where all order flows are Poisson processes:

Market orders are submitted are rate $\lambda_S^M$ with size $\sigma_S$, limit orders are submitted with the same size at rate $\lambda_S^L$ per unit price on a grid with tick size $\Delta P_S$, and all orders are removed randomly with constant probability $\delta_S$ per unit time[1]. Figures 3(a) and 3(b) in Smith et al. (2003) are obtained for different values of a "granularity" parameter $\epsilon_S \propto \frac{\delta_S \sigma_S}{\lambda_S^M}$. It is observed that, when $\epsilon_S$ gets larger, the average book becomes deeper close to the spread, and thinner for higher prices.

Using our own notations, $\epsilon_S$ actually reduces to $\frac{1}{\bar{\lambda}^M}$, i.e. the inverse of the normalized rate of arrival of market orders. Using Smith et al. (2003)'s assumption that limit orders arrive at constant rate $\lambda_S^L$ per unit price and unit time, we obtain in our model $\lambda^L(p,t) = \lambda_S^L$, i.e. $\bar{\Lambda}^L(p) = \lambda_S^L p$. On Fig. 7.1, we plot the average shapes and cumulative shapes of the order book given at Eqs (7.12) and (7.10) with this $\bar{\Lambda}^L$. It turns out that when $\bar{\lambda}^M$ varies, our basic model indeed reproduces precisely Figs 3(a) and 3(b) of Smith et al. (2003).

Therefore we are able to analytically describe the shapes that were only numerically obtained. These shapes can be straightforwardly obtained with different regimes of market orders in our basic model: When the arrival rates of market orders increases (i.e. when $\epsilon_S$ increases), all other things being equal, the average shape of the order book is thinner for lower prices.

### 7.3.2 Empirical and analytical shape in Bouchaud et al. (2002)

We now give two more examples of order book shapes obtained with Eq. (7.12) of Proposition 7.1. We successively consider two types of normalized intensities of arrival rates of limit orders:

- exponentially decreasing with the price: $\bar{\lambda}^L(u) = \frac{\alpha}{\lambda^C} \beta e^{-\beta u}$ ;

- power-law decreasing with the price: $\bar{\lambda}^L(u) = \frac{\alpha}{\lambda^C}(\gamma - 1)(1+u)^{-\gamma}, \gamma > 1$.

The first case is the one observed on Chinese stocks by Gu et al. (2008). The second case is the one suggested in an empirical study by Bouchaud et al. (2002), in which $\gamma \approx 1.5 - 1.7$. Moreover, the latter paper provides the only analytical formula previoulsy proposed (to our knowledge) linking the order flows and the average shape of a limit order book: Bouchaud et al. (2002) derives an analytical formula from a zero-intelligence model by assuming that the price process is diffusive with diffusion constant $D$. Using our notations, their average order book, denoted here $b_{BP}$, is for any $p \in (0, \infty)$:

$$b_{BP}(p) \propto e^{-\sigma^{-1}p} \int_0^p h_{\bar{\lambda}^L}(u) \sinh(\sigma^{-1}u)\, du + \sinh(\sigma^{-1}p) \int_p^\infty h_{\bar{\lambda}^L}(u) e^{-\sigma^{-1}u}\, du, \quad (7.20)$$

---

[1] we have indexed all variables with an $S$ to differentiate them from our own notations

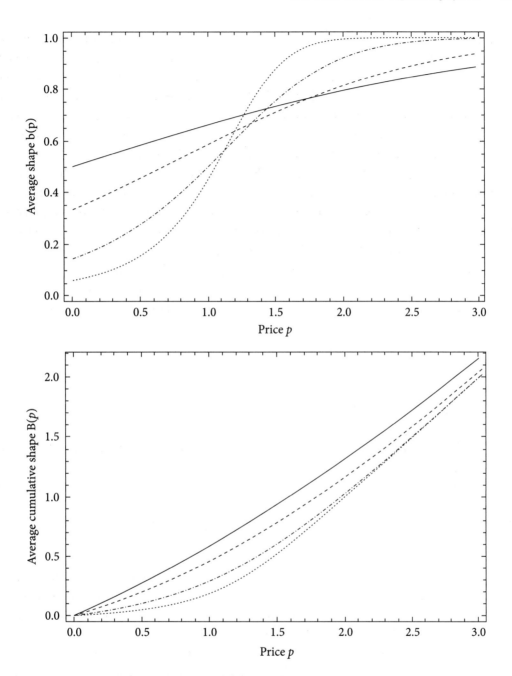

**Fig. 7.1** Shape (top panel) and cumulative shape (bottom panel) of the order book computed using Eqs (7.12) and (7.10) with $\bar{\lambda}^L(p) = \alpha$, $\alpha = 8$ and $\bar{\lambda}^M = 1$ (full line), $\bar{\lambda}^M = 2$ (dashed), $\bar{\lambda}^M = 6$ (dotdashed), $\bar{\lambda}^M = 16$ (dotted). Note that results are scaled on the same dimensionless axes used in Smith et al. (2003). Previously published in Muni Toke (2015)

where the parameter $\sigma^2$ is homogeneous to the variance of a price (it is proportional to the diffusion coefficient divided by the cancellation rate). $\sigma$ is thus interpreted in Bouchaud et al. (2002) as "the typical variation of price during the lifetime of an order, and [it] fixes the scale over which the order book varies". Therefore, although $\sigma$ is not available in our model, since, our one-sided model does not have a diffusive price, we may however obtain a satisfying order of magnitude for the parameter by computing the standard deviation of the price in our model, using numerical simulations (see also Remark 7.3 below).

We plot the shape of Bouchaud et al. (2002) for the two types of normalized arrival rates of limit orders previously mentionned. Note that the formula (7.20) is defined up to a multiplicative constant that we arbitrarily fix such that the maximum offered with respect to the price in our model is equal to the maximum of Eq. (7.20). Results are plotted in Fig. 7.2, and numerical values given in caption.

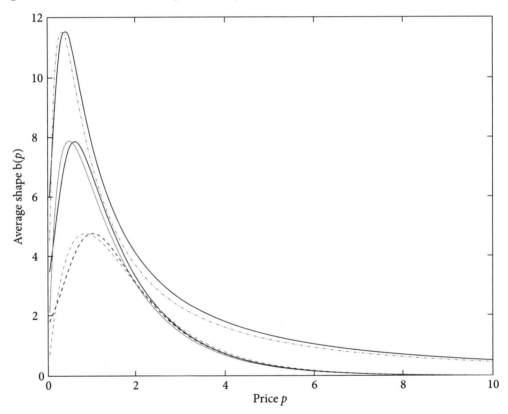

**Fig. 7.2** Comparison of the shapes of the order book in our model (black curves) and using the formula proposed by Bouchaud et al. (2002) (grey curves). Three examples are plotted: Arrival of limit orders with exponential prices, $\alpha = 20$, $\beta = 0.75$, $\lambda^M = 4$, $\lambda^C = 1$ (full lines); idem with $\lambda^M = 8$ (dashed lines); arrival of limit orders with power-law prices, $\alpha = 40$, $\gamma = 1.6$, $\lambda^M = 4$, $\lambda^C = 1$ (dash-dotted lines). Previously published in Muni Toke (2015)

It turns out that our model Eq. (7.12) and Eq. (7.20) provide similar order book shapes. Since, Eq. (7.20) has been successfully tested with empirical data in Bouchaud et al. (2002), Fig. 7.2 provides a good hint that the shape Eq. (7.12) could provide good empirical fittings as well. As $p \to \infty$, both formulas lead to a shape $a(p)$ decreasing as the arrival rate of limit orders $\bar{\lambda}^L(p)$, which was already observed in Bouchaud et al. (2002), and discussed here in Section 7.2. The main difference between the shapes occurs as $p \to 0$. Equation (7.20) imposes that $b_{BP} = 0$, whereas the result Eq. (7.12) allows a more flexible behaviour with $b(0) = \frac{\bar{\lambda}^L(0)}{1+\bar{\lambda}^M}$, a quantity that depends on the three types of order flows. The difference of behaviour close to the spread is not surprising considering the different natures of both models.

**Remark 7.3** We have used numerical simulations of our model to compute the standard deviation of the price in our model. Note however that this could be found by a numerical evaluation of the analytical form of the standard deviation of the price. Indeed, the stationary distribution of the price in our continuous model can be explicitly derived. Assuming this distribution admits a density function $\pi_{PA}$, then the same observation that leads to Eqs (7.6) and (7.7) in the discrete case gives here for any $p \in [0, \infty)$:

$$\int_0^p \pi_{PA}(p)\, dp = 1 - \frac{e^{-\bar{\Lambda}^L(p)}[\bar{\Lambda}^L(p)]^{\bar{\lambda}^M}}{\bar{\lambda}^M \Gamma_{\bar{\Lambda}^L(p)}(\bar{\lambda}^M)}, \tag{7.21}$$

which is written by straighforward differentiation and some terms rearrangements:

$$\pi_{PA}(p) = \bar{\lambda}^L(p)(g_{\bar{\lambda}^M} \circ \bar{\Lambda}^L)(p)\left(1 - \bar{\lambda}^M[\bar{\Lambda}^L(p)]^{-1}(1 - (g_{\bar{\lambda}^M} \circ \bar{\Lambda}^L)(p))\right) \tag{7.22}$$

$$= \bar{\lambda}^L(p) \frac{1 - C(p)}{\bar{\lambda}^M}. \tag{7.23}$$

Some examples of this distribution are plotted on Fig. 7.3.

Now, using this explicit distribution of the price in our order book model, we may compute its standard deviation in the example cases described above, by numerically evaluating the integrals defining the first two moments of the distribution.

Finally, let us recall that the price process in our model is a jump process with right-continuous paths, so it is not diffusive. Note however that the price process in similar two-sided order book models has been shown to admit a diffusive limit with an appropriate time scaling (see Chapter 6).

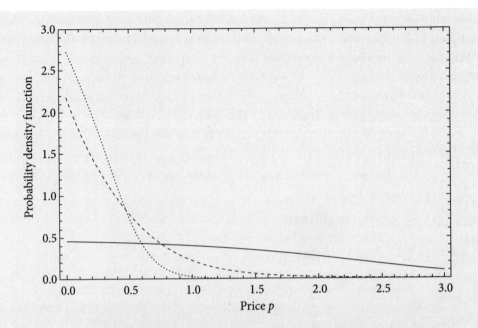

**Fig. 7.3** Price density function $\pi_{p^A}$ as a function of the price, computed with Eq. (7.22), with $h_{\lambda^L}$ constant (full line), exponentially decreasing (dotted) and power-law decreasing (dashed). Previously published in Muni Toke (2015)

## 7.4 A Model with Varying Sizes of Limit Orders

We now allow for random sizes of limit orders in our model. As in Section 7.2, we start by describing the basic model as a queueing system, and then extend it to the case of a continuous price.

Let us recall that we deal with a one-sided order book model, i.e. a model in which all limit orders are ask orders, and all market orders are bid orders. Let $P^A(t)$ denote the price at time $t$. $\{P^A(t), t \in [0 \in \infty)\}$ is a continuous-time stochastic process with value in the discrete set $\{1, \ldots, K\}$, i.e. the price is given in number of ticks. For all $i \in \{1, \ldots, K\}$, (ask) limit orders at price $i$ are submitted according to a Poisson process with parameter $\lambda_i^L$. These processes as assumed to be mutually independent, so that the number of orders submitted at prices between 1 and $r$ (included) is a Poisson process with parameter $\Lambda_r^L$ defined as $\Lambda_r^L = \sum_{i=1}^{r} \lambda_i^L$.

The contribution of this section is to allow for random sizes of limit orders, instead of having unit-size limit orders as in the basic model of Section 7.2. We assume that all the sizes of limit orders are independent random variables. We also assume that the sizes of limit orders submitted at a given price are identically distributed, but we allow this distribution to vary depending on the price. For a given price $k \in \mathbb{N}^*$, let $g_n^k$, $n \in \mathbb{N}^*$, denote the probability that a limit order at price $k$ is of size $n$. Let $\overline{g^k}$ denote the mean size of a limit order at price $k$, which is assumed to be finite. It is a well known property of the Poisson process to state that the rate of arrival of limit orders of size $n$ at price $i$ is $\lambda_i^L g_n^i$, hence the rate of arrival of limit orders of size $n$ with a price lower or equal to $k$ is $\sum_{i=1}^{k} \lambda_i^L g_n^i$. Similarly, the probability that a limit order with a price lower or equal to $k$ is of size $n$ is $G_n^k = \sum_{i=1}^{k} \frac{\lambda_i^L}{\Lambda_k^L} g_n^i$. Let $\overline{G^k} = \sum_{i=1}^{k} \frac{\lambda_i^L}{\Lambda_k^L} \overline{g^i}$ denote the mean size of a limit order with price up to $k$.

Mechanism for cancellation is unchanged: all limit orders standing in the book may be cancelled. Note however that a limit order is not cancelled all at once, but unit by unit, i.e. share by share (see also Remark 7.6 below). It is assumed that the time intervals between the submission of a limit order and the cancellation of one share of this order form a set of mutually independent random variables identically distributed according to an exponential distribution with parameter $\lambda^C > 0$. Finally, (buy) market orders are submitted at random times according to a Poisson process with parameter $\lambda^M$. All market orders are assumed to be of unit size.

As in Section 7.2, let $\{\mathbf{A}_k(t), t \in [0, \infty)\}$ be the stochastic process representing the number of limit orders at prices $1, \ldots, k$ standing in the order book at time $t$. $\mathbf{A}_k$ is thus the cumulative shape of the order book in our model. It can be viewed as the size of a $M^X/M/1 + M$ queueing system with bulk arrival rate $\Lambda_k^L$, bulk volume distribution $(G_n^k)_{n \in \mathbb{N}^*}$, service rate $\lambda^M$ and reneging rate $\lambda^C$ (see e.g., Chaudhry and Templeton (1983) for queueing systems with bulk arrivals). The infinitesimal generator of the process $\mathbf{A}_k$ is thus written:

$$\begin{pmatrix} -\Lambda_k^L & \Lambda_k^L G_1^k & \Lambda_k^L G_2^k & \Lambda_k^L G_3^k & \Lambda_k^L G_4^k & \cdots \\ \lambda^M + \lambda^C & -(\Lambda_k^L + \lambda^M + \lambda^C) & \Lambda_k^L G_1^k & \Lambda_k^L G_2^k & \Lambda_k^L G_3^k & \cdots \\ 0 & \lambda^M + 2\lambda^C & -(\Lambda_k^L + \lambda^M + 2\lambda^C) & \Lambda_k^L G_1^k & \Lambda_k^L G_2^k & \cdots \\ 0 & 0 & \lambda^M + 3\lambda^C & -(\Lambda_k^L + \lambda^M + 3\lambda^C) & \Lambda_k^L G_1^k & \cdots \\ \vdots & \vdots & & \ddots & \ddots & \ddots \end{pmatrix}. \quad (7.24)$$

The stationary distribution $\pi_{\mathbf{A}_k} = (\pi_{\mathbf{A}_k}(n))_{n \in \mathbb{N}}$ of $\mathbf{A}_k$ hence satisfies the following system of equations:

$$\begin{cases} 0 = -\Lambda_k^L \pi_{\mathbf{A}_k}(0) + (\lambda^M + \lambda^C) \pi_{\mathbf{A}_k}(1), \\ 0 = -(\Lambda_k^L + \lambda^M + n\lambda^C) \pi_{\mathbf{A}_k}(n) + (\lambda^M + (n+1)\lambda^C) \pi_{\mathbf{A}_k}(n+1) \\ \quad + \sum_{i=1}^{n} \Lambda_k^L G_i^k \pi_{\mathbf{A}_k}(n-i), \quad (n \geq 1), \end{cases} \quad (7.25)$$

which can be solved by introducing the probability generating functions. Let $\Phi_{A_k}(z) = \sum_{n\in\mathbb{N}} \pi_{A_k}(n) z^n$ and $\Phi_{G^k}(z) = \sum_{n\in\mathbb{N}^*} G_n^k z^n$. Let us also introduce the normalized parameters

$$\bar{\lambda}^M = \frac{\lambda^M}{\lambda^C} \quad \text{and} \quad \bar{\Lambda}_k^L = \frac{\Lambda_k^L}{\lambda^C}. \tag{7.26}$$

By multiplying the $n$-th line by $z^n$ and summing, the previous system leads to the following differential equation:

$$\frac{d}{dz}\Phi_{A_k}(z) + \left(\frac{\bar{\lambda}^M}{z} - \bar{\Lambda}_k^L \varphi_{G^k}(z)\right)\Phi_{A_k}(z) = \frac{\bar{\lambda}^M}{z}\pi_{A_k}(0), \tag{7.27}$$

where we have set $\varphi_{G^k}(z) = \frac{1-\Phi_{G^k}(z)}{1-z}$. This equation is straightforwardly solved to obtain:

$$\Phi_{A_k}(z) = z^{-\bar{\lambda}^M} \bar{\lambda}^M \pi_{A_k}(0) e^{\bar{\Lambda}_k^L \int_0^z \varphi_{G^k}(u)\,du} \int_0^z v^{\bar{\lambda}^M - 1} e^{-\bar{\Lambda}_k^L \int_0^v \varphi_{G^k}(u)\,du}\,dv, \tag{7.28}$$

and the condition $\Phi_{A_k}(1) = 1$ leads to

$$\pi_{A_k}(0) = \left(\bar{\lambda}^M \int_0^1 v^{\bar{\lambda}^M - 1} e^{\bar{\Lambda}_k^L \int_v^1 \varphi_{G^k}(u)\,du}\,dv\right)^{-1}, \tag{7.29}$$

which by substituting in the general solution gives:

$$\Phi_{A_k}(z) = z^{-\bar{\lambda}^M} \frac{\int_0^z v^{\bar{\lambda}^M - 1} e^{\bar{\Lambda}_k^L \int_v^z \varphi_{G^k}(u)\,du}\,dv}{\int_0^1 v^{\bar{\lambda}^M - 1} e^{\bar{\Lambda}_k^L \int_v^1 \varphi_{G^k}(u)\,du}\,dv}. \tag{7.30}$$

Now, turning back to the differential Eq. (7.27), then taking the limit when $z$ tends increasingly to 1 and using basic properties of probability generating functions ($\lim_{z\to 1, t<1} \Phi_{A_k}(z) = 1$, $\lim_{z\to 1, t<1} \frac{d}{dz}\Phi_{A_k}(z) = \mathbf{E}[A_k]$ and $\lim_{z\to 1, t<1} \varphi_{G^k}(z) = \overline{G^k}$), we obtain the result stated in the following proposition.

**Proposition 7.4** *In the discrete one-sided order book model with Poisson arrival at rate $\lambda^M$ of unit size market orders, Poisson arrival of limit orders with rate $\lambda_k^L$ at price $k$ and random size with distribution $(g_n^k)_{n\in\mathbb{N}^*}$ on $\mathbb{N}^*$, and exponential lifetime of non-executed limit orders with parameters $\lambda^C$, the average cumulative shape of the order book up to price $k$ is given by:*

$$\mathbf{E}[A_k] = \bar{\Lambda}_k^L \overline{G^k} - \bar{\lambda}^M + \left(\int_0^1 v^{\bar{\lambda}^M - 1} e^{\bar{\Lambda}_k^L \int_v^1 \varphi_{G^k}(u)\,du}\,dv\right)^{-1}. \tag{7.31}$$

Note that by taking the sizes of all limit orders to be equal to 1, i.e. by setting $g_1^k = 1$ and $g_n^k = 0, n \geq 2$, Eq. (7.31) reduces to Eq. (7.8) of Section 7.2, as expected.

We now introduce a specification of the model where the sizes of limit orders are geometrically distributed with parameter $q \in (0, 1)$ and independent of the price, i.e. for any price $k \in \mathbb{N}^*$, $g_n^k = (1-q)^{n-1}q$. This specification is empirically founded, as it has been observed that the exponential distribution may be a rough continuous approximation of the distribution of the sizes of limit orders (see Chapter 2). This specification of the volume distribution straightfowardly gives $\varphi_{G^k}(z) = \frac{1}{1-(1-q)z}$ and with some computations we obtain:

$$\mathbf{E}[\mathbf{A}_k] = \frac{\bar{\Lambda}_k^L}{q} - \bar{\lambda}^M + \frac{\bar{\lambda}^M q^{\frac{\bar{\Lambda}_k^L}{1-q}}}{{}_2F_1(\bar{\lambda}^M, \frac{-\bar{\Lambda}_k^L}{1-q}, 1+\bar{\lambda}^M, 1-q)}, \tag{7.32}$$

where ${}_2F_1$ is the ordinary hypergeometric function (see e.g., Seaborn, 1991, Chapter 2).

Now, following the idea presented in Section 7.2, we consider an order book with a continuous price, in which limit orders are submitted according to a spatial Poisson process with intensity $\lambda^L(p, t) = \alpha h_{\lambda^L}(p)$. Recall that $h_{\lambda^L}$ is assumed to be a real non-negative function with positive support, denoting the spatial intensity of arrival rates, i.e. the function such that the number of limit orders submitted in the order book in the price interval $[p_1, p_2]$ is a homogeneous Poisson process with rate $\int_{p_1}^{p_2} \alpha h_{\lambda^L}(u) \, du$. Using notations defined in Section 7.2, the cumulative shape at price $p \in [0, +\infty)$ of this continuous order book is thus:

$$A(p) = \frac{1}{q}\bar{\Lambda}^L(p) - \bar{\lambda}^M + \frac{\bar{\lambda}^M q^{\frac{\bar{\Lambda}^L(p)}{1-q}}}{{}_2F_1(\bar{\lambda}^M, \frac{-\bar{\Lambda}^L(p)}{1-q}, 1+\bar{\lambda}^M, 1-q)}, \tag{7.33}$$

which can be derived to obtain the average shape $a(p)$ of the order book, which we state in the following proposition.

**Proposition 7.5** *In a continuous one-sided order book with homogeneous Poisson arrival of unit-size market orders with intensity $\lambda^M$, spatial Poisson arrival of limit orders intensity $\alpha h_{\lambda^L}(p)$, geometric distribution of the sizes of limit orders with parameter $q$, and exponentially distributed lifetimes of non-executed limit orders with parameter $\lambda^C$, the average shape of the order book a is computed for all $p \in [0, \infty)$ by:*

$$a(p) = \frac{\bar{\lambda}^L(p)}{q} + \frac{d}{dp}\left(\frac{\bar{\lambda}^M q^{\frac{\bar{\Lambda}^L(p)}{1-q}}}{{}_2F_1(\bar{\lambda}^M, \frac{-\bar{\Lambda}^L(p)}{1-q}, 1+\bar{\lambda}^M, 1-q)}\right). \tag{7.34}$$

## 7.5 Influence of the Size of Limit Orders on the Shape of the Order Book

We now use the results of Section 7.4 to investigate the influence of the size of the limit orders on the shape of the order book. Recall that market orders are submitted at rate $\lambda^M$ with size 1, that non-executed limit orders are cancelled share by share after a random time with exponential distribution with parameter $\lambda^C$, and that the distribution of the sizes of limit orders is a geometric distribution with parameter $q$ (i.e. with mean $\frac{1}{q}$). We study in detail the influence of the parameter $q$ on the theoretical shape of the order book.

In a first example, we assume that the normalized intensity of arrival of limit orders $\bar{\lambda}^L$ is constant (i.e. as in Smith et al. 2003) and equal to $\alpha q$. Note that when $q$ varies, the mean total volume $\overline{V}(p)$ of arriving limit orders up to price $p$ per unit time remains constant:

$$\overline{V}(p) = \frac{1}{q} \int_0^p \bar{\lambda}^L(u) \, du = p\alpha. \tag{7.35}$$

In other words, when $q$ decreases, limit orders are submitted with larger sizes in average, but less often, keeping the total submitted volume constant. The first remarkable observation is that, although the mean total volumes of limit and market orders are constant, the shape of the order book varies widely with $q$. On Fig. 7.4, we plot the shape $a(p)$ defined at Eq. (7.34), and cumulative shape defined at Eq. (7.33), of an order book with arriving volumes of limit orders as in Eq. (7.35). With the chosen numerical values, the average volume of one limit order ranges from approximately 1 ($q = 0.99$) to 20 ($q = 0.05$). It appears that when $q$ decreases, the shape of the order book increases for lower prices. In other words, *the larger the size of incoming limit orders, the deeper the order book around the spread*, all other things being equal.

We observe that Fig. 7.4 here is similar to Fig. 7.1 here and figure 3 in Smith et al. (2003). However, volumes of limit and market orders are equal in the two latter cases, and we have shown in Section 7.3 that these different shapes can actually be obtained with different regimes of market orders, but equal sizes of market and limit orders: when the arrival rates of market orders increases, all other things being equal, the average shape of the order book is thinner for lower prices.

Therefore, the observation made now is different. In *similar trading regimes* where the mean total volume of limit and market orders are equal, we highlight the influence of the relative volume of limit orders (compared to unit market orders) on the order book shape: The smaller the average size of limit orders, the shallower the order book close to the spread.

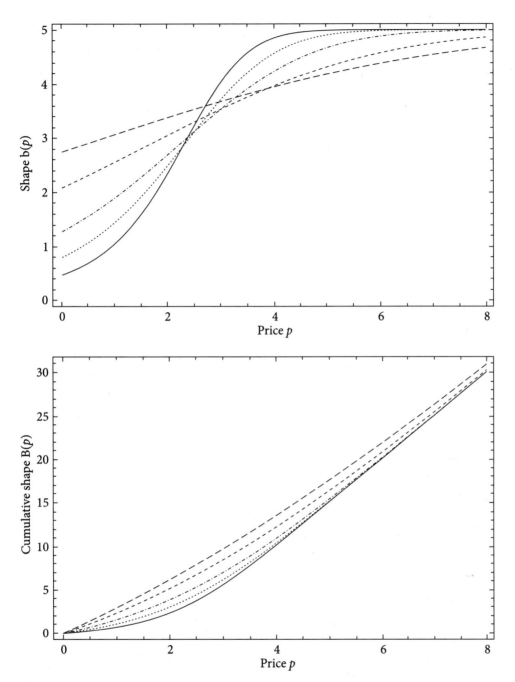

**Fig. 7.4** Shape of the order book as computed in Eq. (7.34) (top) and cumulative shape of the order book as computed in Eq. (7.33) (bottom) with $\bar{\lambda}^M = 10$, $\bar{\lambda}^L(p) = \frac{\alpha q}{K} \mathbf{1}_{(0,K)}$, $\alpha = 40$, $K = 8$, and $q = 0.99$ (full line), $q = 0.5$ (dotted), $q = 0.25$ (dotdashed), $q = 0.10$ (short-dashed), $q = 0.05$ (long-dashed). Previously published in Muni Toke (2015)

We provide a second example of the phenomenon by assuming that the intensity of incoming limit orders exhibits a power-law decrease with the price, as tested in Section 7.3 to compare to the analytical shape provided in Bouchaud et al. (2002). We thus have now $\bar{\lambda}^L(p) = q\alpha(\gamma - 1)(1 + p)^{-\gamma}$. There again, when $q$ varies, the average total volume of incoming limit orders up to price $p$ remains constant and equal to $\alpha(\gamma - 1)\int_0^p (1 + u)^{-\gamma} du = \alpha(1 - (1 + p)^{1-\gamma})$. Figure 7.5 plots the shape of the order book with these characteristics, when $q$ varies.

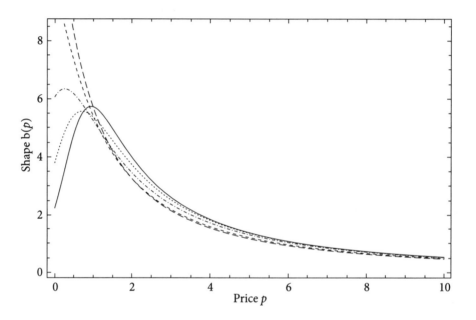

**Fig. 7.5** Shape of the order book as computed in Eq. (7.34) with $\bar{\lambda}^M = 10$, $\bar{\lambda}^L(p) = q\alpha(\gamma - 1)(1 + p)^{-\gamma}$, $\alpha = 40$ $\gamma = 1.6$, and $q = 0.99$ (full line), $q = 0.5$ (dotted), $q = 0.25$ (dotdashed), $q = 0.10$ (short-dashed), $q = 0.05$ (long-dashed). Previously published in Muni Toke (2015)

The observed phenomenon is equally clear with this more realistic distribution of incoming limit orders: the order book deepens at the first limits when the average volume of limit orders increases, the total volume of limit orders and unit-size market orders submitted being constant.

Finally, one might remark that, by assuming unit-size market orders, our model with geometric distribution of limit orders' sizes does not predict the order book shape in the case where the average size of limit orders is smaller than the average size of market orders. In fact, it will now appear that this case has never been empirically encountered in our data: The average size of limit orders is always in our sample greater than the average size of market orders.

**Remark 7.6** We recall that, for analytical tractability purposes, the cancellation mechanism used in this chapter is that each share of submitted limit orders has a lifetime following an exponential distribution independent of the other variables. A more realistic cancellation mechanism would be that each limit order has an independent exponentially-distributed lifetime, i.e. that an order standing in the book is cancelled all at once and not share by share. It turns out this does not qualitatively change our results. We simulate an order book with this latter cancellation system, and compare the average shape of the simulated order book with our analytical formulas. Results are shown on Fig. 7.6 in the case of a uniform price distribution for the limit orders.

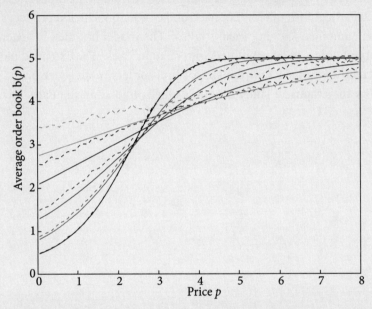

**Fig. 7.6** Shape of the order book with share-by-share cancellation (Eq. (7.34), full lines), and with order-by-order cancellation (numerical simulation, dashed lines). In these computations, $\lambda^M = 10$, $\lambda^C = 1$, $\bar{\lambda}^L(p) = \frac{q\alpha}{\lambda^C K}$, $\alpha = 40$, $K = 8$, and (from black to light grey) $q = 0.99$, $q = 0.5$, $q = 0.25$, $q = 0.10$, $q = 0.05$. Previously published in Muni Toke (2015).

Obviously, if $q$ is close to 1, then both cancellation mechanisms are equivalent and the simulated shape converges to our analytical results. When $q \neq 1$, we cannot straightforwardly compare the two models (the cancellation rate is proportional to the number of shares in the book in one case, and to the number of orders in the book in the other one), but we nevertheless observe that both mechanisms exhibit similar shapes of the order book. Furthermore, it is interesting to note that, irrespective of the cancellation mechanism, our observation stating that the submission of larger limit orders leads to thicker

> order books around the spread is always valid. We may finally add that since, market orders are unit-sized, $q$ may be understood as a relative measure of the volume of limit orders compared to the volume of market orders, and is therefore expected to be closer to 1 than to 0.

## 7.6 Conclusion

We have presented in this chapter the analysis of a simple limit order book model based on classical results from queueing theory. A continuous version of the model has been derived, providing an analytical formula for the shape of the order book that reproduces results from acknowledged numerical and empirical studies. The model has then be extended so as to allow for the limit orders to be submitted with random sizes. The extended model provides some insight into the influence of the size of limit orders in an order book. This insight is confirmed by the empirical study on liquid stocks traded on the Paris Stock Exchange presented in Chapter 3.

# CHAPTER 8

# Advanced Modelling of Limit Order Books

## 8.1 Introduction

This chapter is devoted to the study of more realistic models where the interactions between agents - or, from a statistical point of view, the dependencies between orders of different types - are incorporated in the models. After reviewing in the first section the existing literature on *agent – based modelling* of market interactions, we introduce and analyze in the spirit of Chapter 6, a Hawkes process-based limit order book model.

## 8.2 Towards Non-trivial Behaviours: Modelling Market Interactions

In early statistical models of limit order books, flows of orders are treated as independent processes. Even if the process is empirically detailed and not trivial [Mike and Farmer (2008)], the standing assumption is that order arrivals are independent and identically distributed. This very strong (and false) hypothesis is similar to the "representative agent" hypothesis in Economics: orders being successively and independently submitted, we may not expect anything but regular behaviours. Following the work of economists such as Kirman (1992, 1993, 2002), one has to translate the heterogeneous property of the markets into the agent-based models. Agents are not identical, and not independent.

In this section we present some toy models implementing mechanisms that aim at bringing heterogeneity: Herding behaviour on markets in Cont and Bouchaud (2000), trend following behaviour in Lux and Marchesi (2000) or in Preis et al. (2007), threshold behaviour Cont (2007). Most of the models reviewed in this section are not order book models, since a persistent order book is not kept during the simulations. They are rather price models, where the price changes are determined by the aggregation of excess supply and demand. However, they identify essential mechanisms that may explain some empirical data, and lay the grounds for the designs of better limit order book models.

## 8.2.1 Herding behaviour

The model presented in Cont and Bouchaud (2000) considers a market with $N$ agents trading a given stock with price $p(t)$. At each time step, agents choose to buy or sell one unit of stock, i.e. their demand is $\phi_i(t) = \pm 1, i = 1, \ldots, N$ with probability $a$ or are idle with probability $1 - 2a$. The price change is assumed to be linearly linked with the excess demand $D(t) = \sum_{i=1}^{N} \phi_i(t)$ with a factor $\lambda$ measuring the liquidity of the market:

$$p(t+1) = p(t) + \frac{1}{\lambda} \sum_{i=1}^{N} \phi_i(t). \tag{8.1}$$

$\lambda$ can also be interpreted as a market depth, i.e. the excess demand needed to move the price by one unit. In order to evaluate the distribution of stock returns from Eq. (8.1), we need to know the joint distribution of the individual demands $(\phi_i(t))_{1 \leq i \leq N}$. If the distribution of the demand $\phi_i$ is independent and identically distributed with finite variance, then the Central Limit Theorem stands and the distribution of the price variation $\Delta p(t) = p(t+1) - p(t)$ will converge to a Gaussian distribution as $N$ goes to infinity.

The idea to obtain non-trivial behaviours is to model the diffusion of the information among traders by randomly linking their demand through clusters. At each time step, agents $i$ and $j$ can be linked with probability $p_{ij} = p = \frac{c}{N}$, $c$ being a parameter measuring the degree of clustering among agents. Therefore, an agent is linked to an average number of $(N-1)p$ other traders. Once clusters are determined, the demand are forced to be identical among all members of a given cluster. Denoting $n_c(t)$ the number of cluster at a given time step $t$, $W_k$ the size of the $k$-th cluster, $k = 1, \ldots, n_c(t)$ and $\phi_k = \pm 1$ its investement decision, the price variation is then straightforwardly written:

$$\Delta p(t) = \frac{1}{\lambda} \sum_{k=1}^{n_c(t)} W_k \phi_k \tag{8.2}$$

This modelling is a direct application to the field of finance of the random graph framework as studied in Erdos and Renyi (1960). Kirman (1983) previously suggested it in economics. Using these previous theoretical works, and assuming that the size of a cluster $W_k$ and the decision taken by its members $\phi_k(t)$ are independent, the author are able to show that the distribution of the price variation at time $t$ is the sum of $n_c(t)$ independent identically distributed random variables with heavy-tailed distributions:

$$\Delta p(t) = \frac{1}{\lambda} \sum_{k=1}^{n_c(t)} X_k, \tag{8.3}$$

where the density $f(x)$ of $X_k = W_k \phi_k$ is decaying as:

$$f(x) \underset{|x| \to \infty}{\sim} \frac{A}{|x|^{5/2}} e^{\frac{-(c-1)|x|}{W_0}}. \tag{8.4}$$

Thus, this simple toy model exhibits fat tails in the distribution of prices variations, with a decay reasonably close to empirical data. Therefore, Cont and Bouchaud (2000) show that taking into account a naive mechanism of communication between agents (herding behaviour) is able to drive the model out of the Gaussian convergence and produce non-trivial shapes of distributions of price returns.

### 8.2.2 Fundamentalists and trend followers

Lux and Marchesi (2000) proposed a model very much in line with agent-based models in behavioural finance, but where trading rules are kept simple enough so that they can be identified with a presumably realistic behaviour of agents. This model considers a market with $N$ agents that can be part of two distinct groups of traders: $n_f$ traders are "fundamentalists", who share an exogenous idea $p_f$ of the value of the current price $p$; and $n_c$ traders are "chartists" (or trend followers), who make assumptions on the price evolution based on the observed trend (mobile average). The total number of agents is constant, so that $n_f + n_c = N$ at any time. At each time step, the price can be moved up or down with a fixed jump size of one tick. The probability to go up or down is directly linked to the excess demand $ED$ through a coefficient $\beta$. The demand of each group of agents is determined as follows:

- Each fundamentalist trades a volume $V_f$ proportional, with a coefficient $\gamma$, to the deviation of the current price $p$ from the perceived fundamental value $p_f$: $V_f = \gamma(p_f - p)$.
- Each chartist trades a constant volume $V_c$. Denoting $n_+$ the number of optimistic (buyer) chartists and $n_-$ the number of pessimistic (seller) chartists, the excess demand by the whole group of chartists is written $(n_+ - n_-)V_c$.

Therefore, assuming that there exists some noise traders on the market with random demand $\mu$, the global excess demand is written:

$$ED = (n_+ - n_-)V_c + n_f \gamma(p_f - p) + \mu. \tag{8.5}$$

The probability that the price goes up (resp. down) is then defined to be the positive (resp. negative) part of $\beta ED$.

As observed in Wyart and Bouchaud (2007), fundamentalists are expected to stabilize the market, while chartists should destabilize it. In addition, following Cont and Bouchaud (2000), the authors expect non-trivial features of the price series to results from herding behaviour and transitions between groups of traders. Referring to Kirman's work as well, a mimicking behaviour among chartists is thus proposed. The $n_c$ chartists can change their view on the market (optimistic, pessimistic), their decision being based on a clustering process modelled by an opinion index $x = \frac{n_+ - n_-}{n_c}$ representing the weight of the majority. The probabilities $\pi_+$ and $\pi_-$ to switch from one group to another are formally written:

$$\pi_\pm = v \frac{n_c}{N} e^{\pm U}, \qquad U = \alpha_1 x + \alpha_2 p/v, \tag{8.6}$$

where $v$ is a constant, and $\alpha_1$ and $\alpha_2$ respectively reflect the weight of the majority's opinion and the weight of the observed price in the chartists' decision. Transitions between fundamentalists and chartists are also allowed, decided by comparison of expected returns (see Lux and Marchesi (2000) for details).

The authors show that the distribution of returns generated by their model have excess kurtosis. Using a Hill estimator, they fit a power law to the fat tails of the distribution and observe exponents grossly ranging from 1.9 to 4.6. They also check hints for volatility clustering: absolute returns and squared returns exhibit a slow decay of autocorrelation, while raw returns do not. It thus appears that such a model can grossly fit some stylized facts. However, the number of parameters involved, as well as the complicated rules of transition between agents, make clear identification of sources of phenomenons and calibration to market data difficult and intractable.

Alfi et al. (2009a,b) provide a somewhat simplifying view on the Lux-Marchesi model. They clearly identify the fundamentalist behaviour, the chartist behaviour, the herding effect and the observation of the price by the agents as four essential effects of an agent-based financial model. They show that the number of agents plays a crucial role in a Lux-Marchesi-type model: more precisely, the stylized facts are reproduced only with a finite number of agents, not when the number of agents grows asymptotically, in which case the model stays in a fundamentalist regime. There is a finite-size effect that may prove important for further studies.

The role of the trend following mechanism in producing non-trivial features in price time series is also studied in Preis et al. (2007). The starting point is an order book model similar to Challet and Stinchcombe (2001) or Smith et al. (2003): At each time step, liquidity providers submit limit orders at rate $\lambda$ and liquidity takers submit market orders at rate $\mu$. As expected, this zero-intelligence framework does not produce fat tails in the distribution of (log-)returns nor an over-diffusive Hurst exponent. Then, a stochastic link between order placement and market trend is added: It is assumed that liquidity providers observing a trend in the market will act consequently and submit limit orders at a wider depth in the order book. Although the assumption behind such a mechanism may not be empirically confirmed (a questionable symmetry in order placement is assumed) and should be further discussed, it is interesting enough that it directly provides fat tails in the log-return distributions and an over-diffusive Hurst exponent $H \approx 0.6 - 0.7$ for medium time-scales.

### 8.2.3 Threshold behaviour

We finally review a model focusing primarily on reproducing the stylized fact of volatility clustering, while most of the previous models we have reviewed were mostly focused on fat tails of log returns. Cont (2007) proposes a model with a rather simple mechanism to create volatility clustering. The idea is that volatility clustering characterizes several regimes of volatility (quite periods vs bursts of activity). Instead of implementing an exogenous change of regime, the author defines the following trading rules.

At each period, an agent $i \in \{1, \ldots, N\}$ can issue a buy or a sell order: $\phi_i(t) = \pm 1$. Information is represented by a series of i.i.d Gaussian random variables. $(\epsilon_t)$. This public information $\epsilon_t$ is a forecast for the value $r_{t+1}$ of the return of the stock. Each agent $i \in \{1, \ldots, N\}$ decides whether to follow this information according to a threshold $\theta_i > 0$ representing its sensibility to the public information:

$$\phi_i(t) = \begin{cases} 1 & \text{if } \epsilon_i(t) > \theta_i(t) \\ 0 & \text{if } |\epsilon_i(t)| < \theta_i(t) \\ -1 & \text{if } \epsilon_i(t) < -\theta_i(t) \end{cases} \qquad (8.7)$$

Then, once every choice is made, the price evolves according to the excess demand $D(t) = \sum_{i=1}^{N} \phi_i(t)$, in a way similar to Cont and Bouchaud (2000). At the end of each time step $t$, threshold are asynchronously updated. Each agent has a probability $s$ to update its threshold $\theta_i(t)$. In such a case, the new threshold $\theta_i(t+1)$ is defined to be the absolute value $|r_t|$ of the return just observed. In short:

$$\theta_i(t+1) = \mathbf{1}_{\{u_i(t)<s\}} |r_t| + \mathbf{1}_{\{u_i(t)>s\}} \theta_i(t). \qquad (8.8)$$

The author shows that the time series simulated with such a model do exhibit some realistic facts on volatility. In particular, long range correlations of absolute returns is observed. The strength of this model is that it directly links the state of the market with the decision of the trader. Such a feedback mechanism is essential in order to obtain non trivial characteristics. Of course, the model presented in Cont (2007) is too simple to be fully calibrated on empirical data, but its mechanism could be used in a more elaborate agent-based model in order to reproduce the empirical evidence of volatility clustering.

### 8.2.4 Enhancing zero-intelligence models

We will show in Section 8.3 that the zero-intelligence framework of Chapters 6 and 7 can be generalized to the case of non-Poissonian orders flows. We end this survey section by mentioning very recent developments built on this trend of modelling.

Huang et al. (2015) propose a Markovian order book model in which the intensities of arrival of orders are state-dependent. The order book is represented as a collection of queues indexed by their distance in ticks to a reference price. This reference price process is the main difference with the models described in this book. For each of the queues of the limit order book, market orders, limit orders and cancellation orders are submitted with intensities that are function of the number of shares standing in the book at the time of submission. The authors do not assume any parametric form for these intensities, but plug empirical estimates in their simulations. They show that such a small improvement (the dependency of the intensities on the size of the queue) can lead to a realistic modelling of the stationary state of the order book (distribution of the volume at the best quote). In an extended version of the model, the empirically estimated intensities also depend on the size of the preceding queues, according to their classifications as empty, low, normal or high. In this setting, the authors can simulate quantities of interests, such as the probability of execution of a limit order. Very recently, in the spirit of what we will present in the rest of the chapter, Huang and Rosenbaum (2015) show that ergodicity and diffusive limit of the price process are also obtained in such a setting, with a reference price and state-dependent order flows.

Finally, the use of Hawkes processes for modelling limit order books is a very active and fruitful direction of research, it will be studied in depth in the next sections of this chapter and in Chapter 9.

## 8.3   Limit Order Book Driven by Hawkes Processes

Hawkes processes are a class of point processes that offer very natural and flexible models for processes that mutually excite one another. Since, their introduction, they have been applied to a wide range of research areas, from seismology in the pioneering work Hawkes (1971) to credit risk, financial contagion and more recently, to the modelling of market microstructure. Among the growing litterature in this latter field, Bacry et al. (2013a, 2012, 2013b) or Da Fonseca and Zaatour (2014b,a) introduce and study models where the joint price and order flow dynamics are driven by Hawkes processes. In the recentBacry et al. (2014), the authors develop a new method to accurately estimate non-parametric slowly decaying Hawkes kernels, that allow to describe significant inter-order excitation over long time windows. They fit an eight-dimensional Hawkes model with this type of kernels (four types of orders per side of the book, namely orders that move the price, then market, limit and cancellations that do not move the price), and confirm the self-excitation of the order flows we describe and model in Chapters 4 and 9. Several recent papers [Hardiman et al. (2013); Filimonov and Sornette (2015); Lallouache and Challet (2015) Gatheral et al. (2015)] are concerned with the stability of Hawkes processes calibrated to price dynamics, whereas Alfonsi and Blanc (2015) addresses the optimal execution strategies when the market orders are modelled *via* Hawkes processes.

Closer in spirit to our approach and motivations, the pioneering work by Large (2007) is concerned with the specification, and calibration on real data, of a Hawkes process-based model of limit order books. Muni Toke (2011); Muni Toke and Pomponio (2011) are empirical and numerical studies of Hawkes processes modelling limit order books, and Zheng et al. (2014) is a stylized order book model model driven by Hawkes processes.

As it turns out, the relevance of Hawkes processes for limit order book modelling is amply demonstrated by several empirical properties of the order flow of market and limit orders at the microscopic level. In particular, Hawkes processes exhibit the property of *time clustering*, which can reproduce the fact that order arrivals alternate bursting and quiet periods, as illustrated in Chapter 2. Hawkes processes also exhibit the property of *mutual excitation*, which can reproduce the fact that order flows exhibit non-negligible cross-dependencies, as illustrated in Chapter 4.

The rest of this chapter is devoted to the study of Hawkes process-based limit order book models in a Markovian setting. After describing the mathematical framework, the emphasis will be set, as in Chapter 6, on the ergodicity of the limit order book and the diffusive behaviour of the price at large time scales.

### 8.3.1 Hawkes processes

We briefly recall in this section several classical results on multivariate Markovian Hawkes processes.

Let $\mathbf{N} = (N^1, ..., N^D)$ be a $D$-dimensional point process with intensity vector $\lambda = (\lambda^1, ..., \lambda^D)$.

> **Definition 8.1** We say that $\mathbf{N} = (N^1, ..., N^D)$ is a multivariate Hawkes process with exponential kernel if there exists $(\lambda_0^i)_{1 \leq i \leq D} \in (\mathbb{R}_+^*)^D$, $(\alpha_{ij})_{1 \leq i,j \leq D} \in (\mathbb{R}_+^*)^{D^2}$ and $(\beta_{ij})_{1 \leq i,j \leq D} \in (\mathbb{R}_+^*)^{D^2}$ such that the intensities satisfy the following set of relations:
> 
> $$\lambda^m(t) = \lambda_0^m + \sum_{j=1}^{D} \alpha_{mj} \int_0^t e^{-\beta_{mj}(t-s)} dN^j(s) \qquad (8.9)$$
> 
> for $1 \leq m \leq D$.

The particular choice of exponential kernels is motivated by an important result that we now recall.

**Proposition 8.2** *Define the processes $\mu^{ij}$ as*

$$\mu^{ij}(t) = \alpha_{ij} \int_0^t e^{-\beta_{ij}(t-s)} dN^j(s), \ 1 \le i, j \le D,$$

*and let $\mu = \{\mu^{ij}\}_{1 \le i,j \le D}$. Then, the process $(\mathbf{N}, \mu)$ is Markovian.*

**proof** Lemma 6 in Massoulié (1998) gives a proof of this result.

### Stationarity

Extending the early stability and stationarity result in Hawkes and Oakes (1974), Theorem 5 in Massoulié (1998) proves a general stability result for the multivariate Hawkes processes just introduced. In fact, one can show the existence of a Lyapunov function for such a process. The existence of a Lyapunov function actually implies exponential convergence towards the stationary distribution, a property already seen and used in Chapter 6 (see Appendix C.1.1 for details).

We summarize these results in the following proposition:

**Proposition 8.3** *Let the matrix $\mathbf{A}$ be defined by*

$$\mathbf{A}_{ij} = \frac{\alpha_{ji}}{\beta_{ji}}, \quad 1 \le i, j \le D.$$

*Assume that $\mathbf{A}$ is positive and that its spectral radius $\rho(\mathbf{A})$ satisfies the condition*

$$\rho(\mathbf{A}) < 1. \tag{8.10}$$

*Then, there exists a (unique) multivariate point process $\mathbf{N} = (N^1, \ldots, N^m)$ whose intensity is specified as in Definition 8.1. Moreover, this process is stable, and converges exponentially fast in the total variation norm towards its unique stationary distribution.*

Appendix C, Section C.1.1 provides an explicit construction of Lyapunov functions of arbitrary high polynomial growth at infinity for Hawkes processes.

### 8.3.2 Model setup

A limit order book model whose dynamics is governed by Hawkes processes is now introduced. We shall use the same notations and conventions as in Chapter 6 to represent the limit order book.

The same three types of events can modify the limit order book: arrival of a new limit order, arrival of a new market order, cancellation of an already existing limit order. This time, the arrival of market and limit orders are described by *mutually exciting* Hawkes processes:

- $M^{\pm}(t)$: Hawkes processes for buy or sell market orders, with intensities $\lambda^{M^+}$ and $\lambda^{M^-}$;
- $L_i^{\pm}(t)$: Hawkes processes for limit orders at level $i$, with intensities $\lambda_i^{L^{\pm}}$,

whereas the arrival of a cancellation order is modelled as in Chapter 6 by a doubly stochastic Poisson process:

- $C_i^{\pm}(t)$: Counting process for cancellations of limit orders at level $i$, with intensity $\lambda_i^{C^+} a_i$ and $\lambda_i^{C^-} |b_i|$.

### 8.3.3 The infinitesimal generator

A Markovian $(2K+2)$-dimensional Hawkes process now models the intensities of the arrivals of market and limit orders. The full limit order book can be characterized by the $D$-dimensional process $(\mathbf{a}; \mathbf{b}; \mu)$ of the available quantities and the intensities of the Hawkes processes decomposed as in Section 8.3.1, where $D = (2K+2)^2 + 2K$ is the dimension of the state space.

The infinitesimal generator associated with the process describing the joint evolution of the limit order book has the following expression

$$\mathcal{L}F(\mathbf{a}; \mathbf{b}; \mu) = \lambda^{M^+} \left( F\left([a_i - (q - A(i-1))_+]_+; J^{M^+}(\mathbf{b}); \mu + \Delta^{M^+}(\mu)\right) - F \right)$$

$$+ \sum_{i=1}^{K} \lambda_i^{L^+} \left( F\left(a_i + q; J^{L_i^+}(\mathbf{b}); \mu + \Delta^{L_i^+}(\mu)\right) - F \right)$$

$$+ \sum_{i=1}^{K} \lambda_i^{C^+} a_i \left( F\left(a_i - q; J^{C_i^+}(\mathbf{b}); \mu\right) - F \right)$$

$$+ \lambda^{M^-} \left( F\left(J^{M^-}(\mathbf{a}); [b_i + (q - B(i-1))_+]_-; \mu + \Delta^{M^-}(\mu)\right) - F \right)$$

$$+ \sum_{i=1}^{K} \lambda_i^{L^-} \left( F\left(J^{L_i^-}(\mathbf{a}); b_i - q; \mu + \Delta^{L_i^-}(\mu)\right) - F \right)$$

$$+ \sum_{i=1}^{K} \lambda_i^{C^-} |b_i| \left( F\left(J^{C_i^-}(\mathbf{a}); b_i + q; \mu\right) - F \right)$$

$$- \sum_{i,j=1}^{D} \beta_{ij} \mu^{ij} \frac{\partial F}{\partial \mu^{ij}}. \tag{8.11}$$

In order to ease the already cumbersome notations, we have written $F(a_i; \mathbf{b}; \mu)$ instead of $F(a_1, \ldots, a_i, \ldots, a_K; \mathbf{b}; \mu)$, and use the same symbol for a process and the corresponding state variable in the state space. Moreover, the notation $\Delta^{(\cdots)}(\mu)$ stands for the jump of the intensity vector $\mu$ corresponding to a jump of the process $N^{(\cdots)}$ (see Section C.1.1).

The operator $\mathcal{L}$ is a combination of

- standard difference operators corresponding to the arrival or cancellation of orders at each limit and shift operators expressing the moves in the best limits, as already seen;
- drift terms coming from the mean-reverting behaviour of the intensities of the Hawkes processes between jumps.

Note that, similarly to what is done in Chapter 6, the infinitesimal generator is fully worked out in the case of a discrete state space for the quantities $\mathbf{a}, \mathbf{b}$; some trivial but notationally cumbersome modifications would be necessary in order to account for the case of general, real-valued quantities $a_i, b_i$'s and order size $q$.

### 8.3.4 Stability of the order book

In this section, we study the long-time behaviour of the limit order book. A Lyapunov function is built, ensuring the ergodicity of the limit order book under the natural assumption (8.10). More precisely, there holds the following proposition.

**Proposition 8.4** *Under the standing assumptions, in particular (8.10), the limit order book process $\mathbf{X}$ is ergodic. It converges exponentially fast towards its unique stationary distribution $\Pi$.*

**Proof** Given the existence of the Lyapunov function provided in Lemma 8.5 below, the result is proven using Theorem 7.1 in Meyn and Tweedie (1993), see Appendix C, Section C.2.1. The only technical difficulty lies in establishing the fact that compact sets are petite sets, a result proven in Zheng et al. (2014), Theorem 3.1 and Section 3.3.

**Lemma 8.5** *For $\eta > 0$ small enough, the function $V$ defined by*

$$V(\mathbf{a}; \mathbf{b}; \mu) = \sum_{i=1}^{K} a_i + \sum_{i=1}^{K} |b_i| + \frac{1}{\eta} \sum_{i,j=1}^{(2K+2)^2} \delta_{ij} \mu^{ij} \equiv V_1 + \frac{1}{\eta} V_2 \qquad (8.12)$$

*where $V_1$ (resp. $V_2$) corresponds to the part that depends only on $(\mathbf{a}; \mathbf{b})$ (resp. $\mu$), is a Lyapunov function satisfying a geometric drift condition*

$$\mathcal{L}V \leq -\zeta V + C, \qquad (8.13)$$

*for some $\zeta > 0$ and $C \in \mathbb{R}$. The coefficients $\delta_{ij}$'s are defined in (C.13) in Section C.1.1.*

**Proof** First specialize $V_2$ to be identical - up to a change in the indices - to the function defined by Eq. (8.5) in Appendix C.1.1. Regarding the "small" parameter $\eta > 0$, it will become handy as a penalization parameter, as we shall see below.

Thanks to the linearity of $\mathcal{L}$, there holds

$$\mathcal{L}V = \mathcal{L}V_1 + \frac{1}{\eta}\mathcal{L}V_2.$$

The first term $\mathcal{L}V_1$ is dealt with exactly as in Chapter 6:

$$\mathcal{L}V_1 \leq -(\lambda^{M^+} + \lambda^{M^-})q + \sum_{i=1}^{K}\left(\lambda_i^{L^+} + \lambda_i^{L^-}\right)q - \sum_{i=1}^{K}\left(\lambda_i^{C^+}a_i + \lambda_i^{C^-}|b_i|\right)q$$

$$+ \sum_{i=1}^{K}\lambda_i^{L^+}(i_S - i)_+ a_\infty + \sum_{i=1}^{K}\lambda_i^{L^+}(i_S - i)_+|b_\infty| \qquad (8.14)$$

$$\leq -\left(\lambda^{M^+} + \lambda^{M^-}\right)q + \left(\Lambda^{L^-} + \Lambda^{L^+}\right)q - \underline{\lambda^C}qV_1(\mathbf{x})$$

$$+ K\left(\Lambda^{L^-}a_\infty + \Lambda^{L^+}|b_\infty|\right), \qquad (8.15)$$

where

$$\Lambda^{L^\pm} := \sum_{i=1}^{K}\lambda_i^{L^\pm} \text{ and } \underline{\lambda^C} := \min_{1\leq i\leq K}\{\lambda_i^{C^\pm}\} > 0.$$

Computing $\mathcal{L}V_2$ yields an expression identical to that obtained in Section C.1.1:

$$\mathcal{L}(V_2) = \sum_{i,j}\lambda_0^j\delta_{ij}\alpha_{ij} + (\kappa - 1)\sum_{j,k}\epsilon_k\mu^{jk},$$

so that there holds

$$\mathcal{L}V = \mathcal{L}V_1 + \frac{1}{\eta}\mathcal{L}V_2 \leq -\underline{\lambda^C}qV_1 - \frac{\gamma}{\eta}V_2 - \mathbf{G}.\boldsymbol{\mu} + C,$$

where $\gamma$ is as in Eq. (C.15), $\mathbf{G}.\boldsymbol{\mu}$ is a compact notation for the linear form in the $\mu^{ij}$'s obtained in Eq. (8.15), and $C$ is some constant. Now, thanks to the positivity of the coefficients in $V_2$ and of the $\mu^{ij}$'s, one can choose $\eta$ small enough that there holds

$$\forall \mu, \ |G.\mu| \leqslant \frac{\gamma}{2\eta} V_2(\mu),$$

which yields

$$\mathcal{L}V \equiv \mathcal{L}V_1 + \frac{1}{\eta}\mathcal{L}V_2 \leqslant -\underline{\lambda^C} q V_1 - \frac{\gamma}{2\eta} V_2 + C, \tag{8.16}$$

and finally

$$\mathcal{L}V \leqslant -\zeta V + C,$$

with $\zeta = Min\left(\underline{\lambda^C} q, \frac{\gamma}{2\eta}\right)$ and $C$ is some constant.

### 8.3.5 Large scale limit of the price process

Using the same approach as in Chapter 6 in this more general context, we study the long-time behaviour of the price process, taking into account the stochastic behaviour of the intensities of the point processes triggering the order book events. We first recall the expression of the price dynamics in our limit order book model. Consider for instance the mid-price, solution to the SDE (6.29) which we recall here

$$dP(t) = \frac{\Delta P}{2}\left[\left(\mathbf{A}^{-1}(q) - i_S\right)dM^+(t) - \left(\mathbf{B}^{-1}(q) - i_S\right)dM^-(t)\right.$$
$$- \sum_{i=1}^{K}(i_S - i)_+ dL_i^+(t) + \sum_{i=1}^{K}(i_S - i)_+ dL_i^-(t)$$
$$\left. + \left(\mathbf{A}^{-1}(q) - i_S\right)dC_{i_A}^+(t) - \left(\mathbf{B}^{-1}(q) - i_S\right)dC_{i_B}^-(t)\right].$$

Let us recast this equation - exactly as in 6.40 - under the following form:

$$P_t = \int_0^t \sum_i F_i(\mathbf{X}(u)) \, dN^i(u), \tag{8.17}$$

where the $N^i$ are the point processes driving the limit order book, $\mathbf{X}$ is the Markovian process describing its state, and $P$ is one of the price processes we are interested in. In the current context of Hawkes processes, $\mathbf{X} = (\mathbf{a}, \mathbf{b}, \mu)$ and the $N^i$, with state-dependent intensitites $v^i(\mathbf{X})$, are the Poisson and Hawkes processes driving the limit order book. As mentioned earlier, the $F_i$ are bounded functions, as the price changes are bounded by the total number of limits in the book, thanks to the non-zero boundary conditions $a_\infty, b_\infty$.

Denote again by $\Pi$ the stationary distribution of $\mathbf{X}$, as provided by Proposition 8.4. We can prove the following theorem:

**Theorem 8.6** *Write as above the price*

$$P_t = \int_0^t \sum_i F_i(\mathbf{X}(s)) \, dN^i(s)$$

*and its compensator*

$$Q_t = \int_0^t \sum_i v^i(\mathbf{X}(s)) F_i(\mathbf{X}(s)) \, ds.$$

*Define*

$$h = \sum_i v^i F_i(\mathbf{X})$$

*and let*

$$\alpha = \stackrel{a.s.}{\lim_{t \to +\infty}} \frac{1}{t} \int_0^t \sum_i v^i(\mathbf{X}(s)) F_i(\mathbf{X}(s)) \, ds = \int h(\mathbf{X}) \, \Pi(d\mathbf{X}).$$

*Finally, introduce the solution $g$ to the Poisson equation*

$$\mathcal{L}g = h - \alpha$$

*and the associated martingale*

$$Z_t = g(\mathbf{X}(t)) - g(\mathbf{X}_0) - \int_0^t \mathcal{L}g(\mathbf{X}(s)) \, ds \equiv g(\mathbf{X}(t)) - g(\mathbf{X}_0) - Q_t + \alpha t.$$

*Then, the deterministically centered, rescaled price*

$$\bar{P}^n(t) \equiv \frac{P_{nt} - \alpha n t}{\sqrt{n}}$$

*converges in distribution to a Wiener process $\bar{\sigma} W$. The asymptotic volatility $\bar{\sigma}$ satisfies the identity*

$$\bar{\sigma}^2 = \lim_{t \to +\infty} \frac{1}{t} \int_0^t \sum_i v^i(\mathbf{X}(s)) \left( (F_i - \Delta^i(g))(\mathbf{X}(s)) \right)^2 ds \tag{8.18}$$

$$\equiv \int \sum_i v^i(\mathbf{X}) \left( (F_l - \Delta^i(g))(\mathbf{X}) \right)^2 \lambda_i \Pi(d\mathbf{X}). \tag{8.19}$$

**Proof** Using again the martingale method as in the proof of Theorem 6.5 with

$$P_t = (P_t - Q_t) - Z_t + g(\mathbf{X}(t)) - g(\mathbf{X}_0) + \alpha t \equiv (M_t - Z_t) + g(\mathbf{X}(t)) - g(\mathbf{X}_0) + \alpha t,$$

the theorem will be proven if one can show that $g \in L^2(\Pi(d\mathbf{X}))$. The condition

$$h^2 \leqslant V \qquad (8.20)$$

(where $V$ is a Lyapunov function for the process) of Theorem 4.4 in Glynn and Meyn (1996) is sufficient for $g$ to be in $L^2(\Pi(d\mathbf{X}))$. The linear Lyapunov function $V$ introduced in (8.12) does not yield the desired result, because $h$ now has a linear growth. However, Lemma Lemma C.5 in Appendix C provides a Lyapunov function having a polynomial growth of arbitrarily high order in the intensities at infinity, thereby ensuring that Condition(8.20) holds.

## 8.4 Conclusion

The question of modelling the interactions between agents of different types is quite fascinating. It has important consequences on many aspects of the understanding of limit order books, be it from an empirical, theoretical or practical point of view. In this chapter we have suggested and reviewed several, and studied some, avenues for such a refined modelling. It is however clear that much more work is still to be done, in view in particular of the fierce *competition* between agents following different strategies. Such a game-theoretic approach to limit order book modelling is still in its infancy, and will probably be the subject of many interesting future studies.

# PART THREE
# SIMULATION OF LIMIT ORDER BOOKS

# CHAPTER 9

# Numerical Simulation of Limit Order Books

## 9.1 Introduction

This chapter describes useful algorithms and their implementations for the numerical simulation of limit order books. The basic algorithm simulating a zero-intelligence limit order book is presented, and then extended to the case of a multivariate Hawkes process-driven order book. Numerical results are analyzed, and compared to empirical data.

## 9.2 Zero-intelligence Limit Order Book Simulator

### 9.2.1 An algorithm for Poissonian order flows

We describe a basic algorithm for the simulation of the limit order book model of Chapters 6 and 7. We will assume for notational simplicity that the order book is symmetric, i.e. that the intensities of arrival of orders of various types are identical on the bid and ask side. We can thus drop the ± signs of our notations, and define here with obvious notations:

$$\lambda^L = \left(\lambda_1^L, \ldots, \lambda_K^L\right),$$

$$\Lambda^L = \sum_{i=1}^{K} \lambda_i^L,$$

$$\lambda^C(\mathbf{a}) = \left(\lambda_1^C a_1, \ldots, \lambda_K^C a_K\right),$$

$$\Lambda^C(\mathbf{a}) = \sum_{i=1}^{K} \lambda_i^C a_i,$$

$$\lambda^C(\mathbf{b}) = \left(\lambda_1^C |b_1|, \ldots, \lambda_K^C |b_K|\right),$$

$$\Lambda^C(\mathbf{b}) = \sum_{i=1}^{K} \lambda_i^C |b_i|,$$

$$\Lambda(\mathbf{a}, \mathbf{b}) = 2(\lambda^M + \Lambda^L) + \Lambda^C(\mathbf{a}) + \Lambda^C(\mathbf{b}).$$

Using these notations, the routine for the simulation of the limit order book is sketched in Algorithm 1 [see also Gatheral and Oomen (2010) for a similar description].

---

**Algorithm 1**  Order book simulation with Poisson order flows.

---

**Require:** *Model parameters*: $K$ (number of visible limits), $\lambda^M$, $\{\lambda_i^L\}_{i\in\{1,\ldots K\}}$, $\{\lambda_i^C\}_{i\in\{1,\ldots K\}}$ (intensities of order flows), $a_\infty$, $b_\infty$ (size of hidden limits), random distributions $\mathcal{V}_L$, $\mathcal{V}_M$, $\mathcal{V}_C$ (volume of limit, market and cancel orders).

*Simulation Parameters*: $N$ (length of simulation in event time), $\mathbf{X}_{\text{init}}$ (initial state of the limit order book)

1: **Initialization:** Set $t \leftarrow 0$ (physical time), $\mathbf{X}(0) \leftarrow \mathbf{X}_{\text{init}}$.

2: **for** $n = 1, \ldots, N$

3: **Update the cancellation intensities:** $\Lambda^C(\mathbf{b}) = \sum_{i=1}^K \lambda_i^C |b_i|$, $\Lambda^C(\mathbf{a}) = \sum_{i=1}^K \lambda_i^C a_i$.

4: **Time of next event:** Draw the waiting time $\tau$ from an exponential distribution with parameter $\Lambda(\mathbf{a}, \mathbf{b}) = 2(\lambda^M + \Lambda^L) + \Lambda^C(\mathbf{a}) + \Lambda^C(\mathbf{b})$.

5: **Type of next event:** Draw an event type according to the probability vector $\left(\lambda^M, \lambda^M, \Lambda^L, \Lambda^L, \Lambda^C(\mathbf{a}), \Lambda^C(\mathbf{b})\right)/\Lambda(\mathbf{a}, \mathbf{b})$. These probabilities correspond respectively to a buy market order, a sell market order, a buy limit order, a sell limit order, a cancellation of an existing sell order and a cancellation of an existing buy order.

6: **Volume of next event:** Depending on the event type, draw the order volume from one of the random distributions $\mathcal{V}_L$, $\mathcal{V}_M$, $\mathcal{V}_C$.

7: **Price of next event:**

8: **if** the next event is a limit order **then**

9: Draw the relative price level according to the probability vector $\left(\lambda_1^L, \ldots, \lambda_K^L\right)/\Lambda^L$.

10: **end if**

11: **if** the next event is a cancellation **then**

*Contd...*

| **Algorithm 1** Order book simulation with Poisson order flows. |
| --- |
| 12: Draw the relative price level at which to cancel an order from according to the probability vector $\left(\lambda_1^C a_1, \ldots, \lambda_K^C a_K\right)/\Lambda^C(\mathbf{a})$ (ask case) or $\left(\lambda_1^C \|b_1\|, \ldots, \lambda_K^C \|b_K\|\right)/\Lambda^C(\mathbf{b})$ (bid case). |
| 13: **end if** |
| 14: Set $t \leftarrow t + \tau$ and **update the order book** according to the new event. |
| 15: **Enforce the boundary conditions**: $a_i \leftarrow a_\infty, i \geq K+1$ and $b_i \leftarrow b_\infty, i \geq K+1$. |
| 16: **end for** |

This algorithm is simply the transcription of the limit order book modelled in Chapter 6, enhanced as in Chapter 7 to allow random sizes of submitted orders of all types. This feature will help producing more realistic simulated data.

### 9.2.2 Parameter estimation

The parameters of the model are estimated on the dataset presented in Appendix B.6. In this section we analyze the results computed with the parameters estimated for the stock SCHN.PA (Schneider Electric) in March 2011. These results and figures are given as illustration, but it is important to note that they are qualitatively similar for all CAC 40 stocks.

Let $T$ be the length of the time window of interest each day. If $N_T^M$ is the total number of trades (buy and sell) during this time window, then the estimate for the intensity of the market orders is

$$\widehat{\lambda^M} = \frac{N_T^M}{2T}.$$

If $N_{i,T}^L$ is the total number of limit orders (buy and sell) submitted $i$ ticks away from the best opposite quote during the time interval of length $T$, then the estimate for the intensity of the limit orders $i$ ticks away from the best opposite quote is

$$\widehat{\lambda_i^L} = \frac{N_{i,T}^L}{2T}.$$

As for the cancellation intensities, we need to normalize the count by the (temporal) average number of shares $\langle \mathbf{X}_i \rangle$ at distance $i$ from the best opposite quote. If $N_{i,T}^C$ is the total number of cancellation orders (buy and sell) submitted $i$ ticks away from the best opposite quote during the time interval of length $T$, then the estimate for the intensity of the cancellation orders $i$ ticks away from the best opposite quote is

$$\widehat{\lambda_i^C} = \frac{1}{\langle \mathbf{X}_i \rangle} \frac{N_{i,T}^C}{2T}$$

We then average $\widehat{\lambda^M}$, $\widehat{\lambda_i^L}$ and $\widehat{\lambda_i^L}$ across 23 trading days to get the final estimates. As for the volumes, we compute the empirical distributions of the volumes for each type of orders, and we fit by maximum likelihood estimation a log-normal distribution with parameters $(\widehat{v^M}, \widehat{s^M})$ (market orders), $(\widehat{v^L}, \widehat{s^L})$ (limit orders) and $(\widehat{v^C}, \widehat{s^C})$ (cancellation orders).

The parameters estimated for SCHN.PA in March 2011 are summarized in Tables 9.1 and 9.2. A graphic representation of these parameters is given in Fig. 9.1 and Fig. 9.2.

**Table 9.1** Model parameters for the stock SCHN.PA (Schneider Electric) in March 2011 (23 trading days). Figures 9.1 and 9.2 are graphical representation of these parameters

| | |
|---|---|
| K | 30 |
| $a_\infty$ | 250 |
| $b_\infty$ | 250 |
| $(v^M, s^M)$ | (4.00, 1.19) |
| $(v^L, s^L)$ | (4.47, 0.83) |
| $(v^C, s^C)$ | (4.48, 0.82) |
| $\lambda^{M\pm}$ | 0.1237 |

**Table 9.2** Model parameters for the stock SCHN.PA (Schneider Electric) in March 2011 (23 trading days). Figures 9.1 and 9.2 are graphical representation of these parameters

| $i$ (ticks) | $\langle \mathbf{X}_i \rangle$ (shares) | $\lambda_i^{L\pm}$ | $10^3 . \lambda_i^{C\pm}$ |
|---|---|---|---|
| 1 | 276 | 0.2842 | 0.8636 |
| 2 | 1129 | 0.5255 | 0.4635 |
| 3 | 1896 | 0.2971 | 0.1487 |
| 4 | 1924 | 0.2307 | 0.1096 |
| 5 | 1951 | 0.0826 | 0.0402 |
| 6 | 1966 | 0.0682 | 0.0341 |

Contd...

| $i$ (ticks) | $\langle \mathbf{X}_i \rangle$ (shares) | $\lambda_i^{L\pm}$ | $10^3 . \lambda_i^{C\pm}$ |
|---|---|---|---|
| 7 | 1873 | 0.0631 | 0.0311 |
| 8 | 1786 | 0.0481 | 0.0237 |
| 9 | 1752 | 0.0462 | 0.0233 |
| 10 | 1691 | 0.0321 | 0.0178 |
| 11 | 1558 | 0.0178 | 0.0127 |
| 12 | 1435 | 0.0015 | 0.0012 |
| 13 | 1338 | 0.0001 | 0.0001 |
| 14 | 1238 | 0.0 | 0.0 |
| 15 | 1122 | ⋮ | ⋮ |
| 16 | 1036 | | |
| 17 | 943 | | |
| 18 | 850 | | |
| 19 | 796 | | |
| 20 | 716 | | |
| 21 | 667 | | |
| 22 | 621 | | |
| 23 | 560 | | |
| 24 | 490 | | |
| 25 | 443 | | |
| 26 | 400 | | |
| 27 | 357 | | |
| 28 | 317 | | |
| 29 | 285 | ⋮ | ⋮ |
| 30 | 249 | 0.0 | 0.0 |

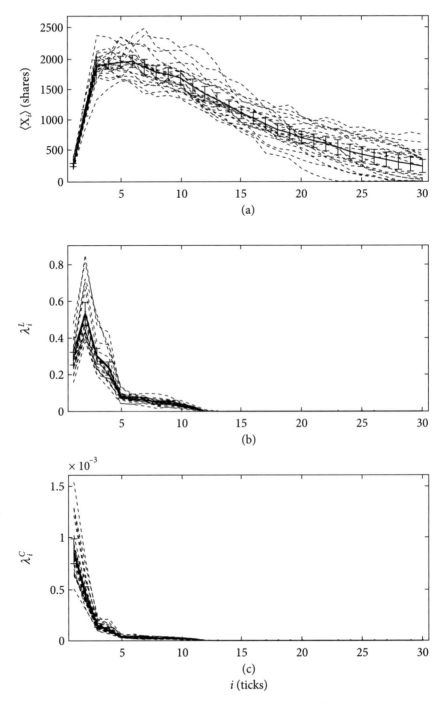

**Fig. 9.1** Model parameters: Arrival rates and average depth profile (parameters as in Table 9.2). Error bars indicate variability across different trading days. Extracted from Abergel and Jedidi (2013)

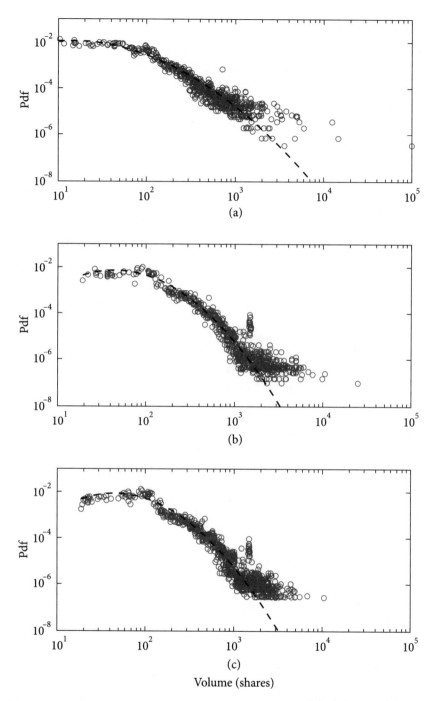

**Fig. 9.2** Model parameters: Volume distribution. Panels $(a), (b)$ and $(c)$ correspond respectively to market, limit and cancellation orders volumes. Dashed lines are lognormal fits (parameters as in Table 9.1). Extracted from Abergel and Jedidi (2013)

120  Limit Order Books

### 9.2.3 Performances of the simulation

We compute on our simulated data several quantities of interest. Figure 9.3 represents the average shape of the order book. Recall that this shape has been analytically determined in Chapter 7 in the case of a one-sided model. The agreement between the simulated shape and the empirical one is fairly good. A cross-sectional view of this quantity for all CAC 40 stocks is provided in the next subsection (Fig. 9.10 Panel $(a)$).

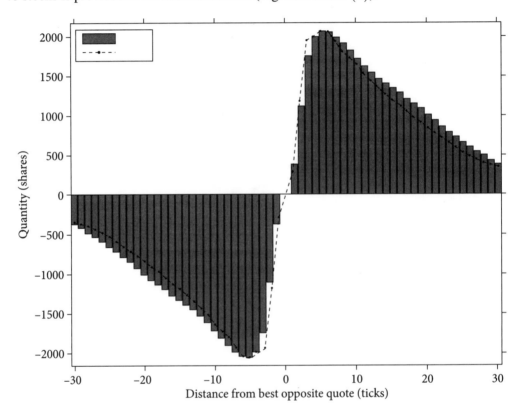

**Fig. 9.3**   Average depth profile. Simulation parameters are summarized in Tables 9.1 and 9.2. Extracted from Abergel and Jedidi (2013)

We also study some properties of the price process derived from the order book simulations. The distribution of the spread is given in Fig. 9.4. We observe that the simulated distribution is *tighter* than the empirical one. This observation stands for all CAC 40 stocks, as documented Fig. 9.10 Panel $(b)$. It must however be taken with a grain of salt, as the spread distribution is highly sensitive to many parameters of the model. In Section 9.3, we present a qualitative study of the spread distribution under various modelling assumptions for the arrival of orders.

Figure 9.5 shows the fast decay of the autocorrelation function of the price increments. Note the high negative autocorrelation of simulated trade prices relatively to the data. This feature is most likely due to the fact that we have assumed a symmetric order book and Poissonian arrival of orders: In real markets, order splitting induces a clustering of market orders of identical signs, so that the traded prices in a sequence of market orders are closer to one another that in the zero-intelligence case for which the *bid-ask bounce* effect[1] is important.

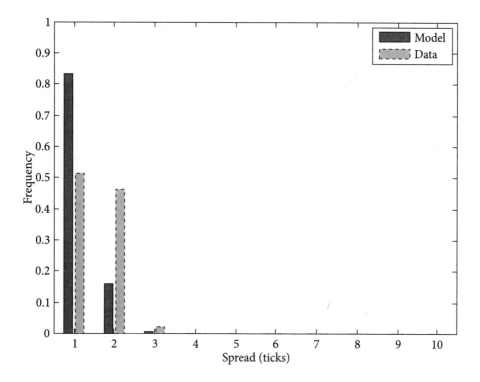

**Fig. 9.4** Probability distribution of the spread. Note that the model (dark gray) predicts a tighter spread than the data. Extracted from Abergel and Jedidi (2013)

Figure 9.6 gives an example of simulated path for the mid-price. Figure 9.7 plots the histogram of the empirical distribution of the price increments over 1000 events. At this (large) scale, the normal distribution is a good match. This is a well-known observation, called asymptotic normality of price increments. Figure 9.8 shows the Q-Q plots of the mid-price increments for four different scales, from 1 second to 5 minutes. The convergence of the distribution of the price increments towards a Gaussian distribution as the time scale of observation increases is clearly observed.

---

[1] The bid-ask bounce effect describes the fact that the signs of market orders generally alternate, thereby creating a large change in traded prices due to the presence of the bid-ask spread

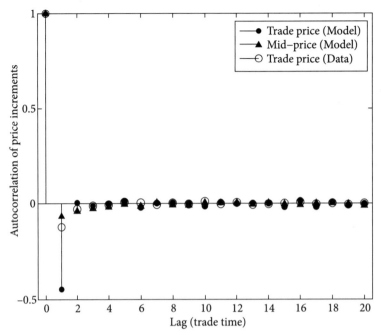

**Fig. 9.5** Autocorrelation of price increments. This figure shows the fast decay of the autocorrelation function, and the large negative autocorrelation of trades at the first lag. Extracted from Abergel and Jedidi (2013)

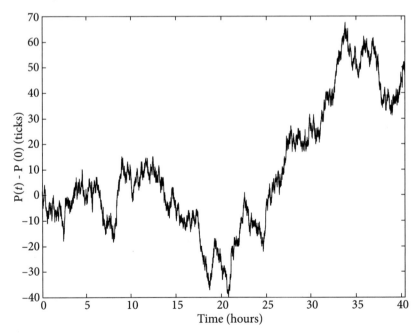

**Fig. 9.6** Price sample path. At large time scales, the price process is close to a Wiener process. Extracted from Abergel and Jedidi (2013)

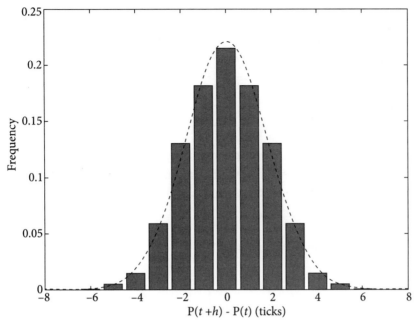

**Fig. 9.7** Probability distribution of price increments. Time lag $h = 1000$ events. Extracted from Abergel and Jedidi (2013)

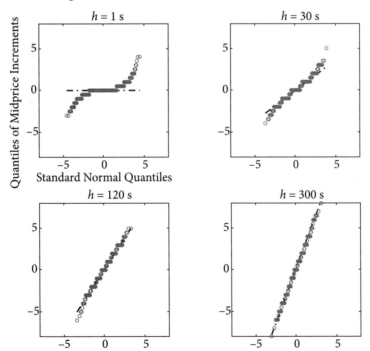

**Fig. 9.8** Q-Q plot of mid-price increments. $h$ is the time lag in seconds. This figure illustrates the aggregational normality of price increments. Extracted from Abergel and Jedidi (2013)

We now give a few facts on the properties of the variance of the price processes of our simulations. The *signature plot* of a price time series is defined as the variance of price increments at lag $h$ normalized by the lag $h$, as a function of this lag $h$. In other words, it is the function $h \mapsto \sigma_h^2$ where

$$\sigma_h^2 = \frac{\mathbf{V}\left[P(t+h) - P(t)\right]}{h}. \tag{9.1}$$

This function measures the variance of price increments per time unit. Its main interest is that it shows the transition from the variance at small time scales where micro-structure effects dominate, to the long-term variance. Using the results of Chapter 6, in particular Theorem 6.5, one can show that

$$\lim_{h \to \infty} \sigma_h^2 = \sigma^2, \text{ for some fixed value } \sigma. \tag{9.2}$$

Figure 9.9 shows the signature plots computed on our simulations compared to the empirical ones. Signature plots are computed for both the trade prices and the mid-prices, and in both event and calendar time.

Two main observations are to be made. First, the simulated long-term variance is lower than the variance computed from the data. This observation remains valid for all CAC 40 stocks as documented in Fig. 9.10 Panel (c). We know that depth (shape) of the order book increases away from the best price towards the center of the book. In the absence of autocorrelation in trade signs, this would cause prices to wander less often far away from the current best as they hit a higher "resistance". We also suspect that actual prices exhibit locally more "drifting phases" than in our symmetric Markovian simulation where the expected price drift is null at all times. An interesting analysis of a simple order book model that allows time-varying arrival rates can be found in Challet and Stinchcombe (2003).

Second, the simulated signature plot is too high at short time scales relative to the asymptotic variance, especially for traded prices. As seen previously, this behaviour is well explained by the bid-ask bounce, which is too strong in the zero-intelligence model as there is no accounting for the clustering of orders of identical signs (see Subsection 9.2.4 below for a simple quantitative analysis of this phenomenon). It is however remarkable that the signature plot of empirical trade prices looks much flatter than the signature plot of simulated trade prices. Indeed, a flat empirical signature plot at all time scales suggests that the prices are actually diffusive, which seems to contradict the observation that empirical order signs exhibit positive long-ranged correlations. This has been observed and discussed in several empirical studies (Bouchaud et al. 2004; Lillo and Farmer, 2004; Farmer et al. 2006; Bouchaud et al. 2009). According to these studies, the paradox is solved by observing that the diffusivity results from two opposite effects:

Numerical Simulation of Limit Order Books 125

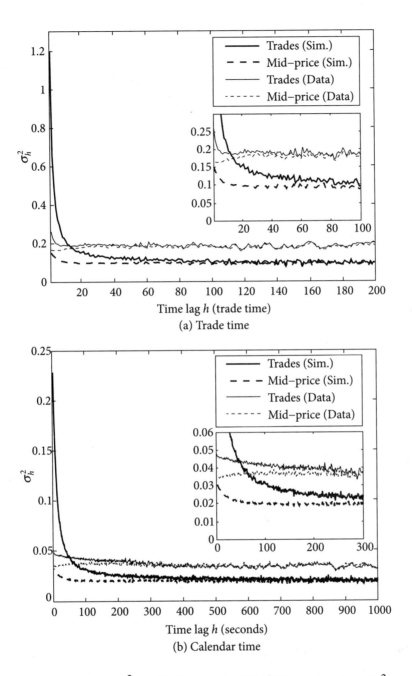

**Fig. 9.9** Signature plot: $\sigma_h^2 := \mathbb{V}\left[P(t+h) - P(t)\right]/h$. y axis unit is tick$^2$ per trade for panel $(a)$ and tick$^2 \cdot$ second$^{-1}$ for panel $(b)$. We used a 1,000,000 event simulation run for the model signature plots. Data signature plots are computed separately for each trading day $[9:30\text{--}14:00]$ then averaged across 23 days. For calendar time signature plots, prices are sampled every second using the last tick rule. The inset is a zoom-in. Extracted from Abergel and Jedidi (2013)

**126** *Limit Order Books*

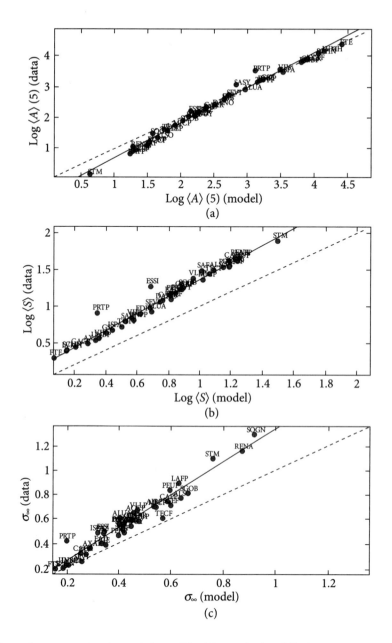

**Fig. 9.10** A cross-sectional comparison of liquidity and price diffusion characteristics between the model and data for CAC 40 stocks (March 2011). Extracted from Abergel and Jedidi (2013)

On the one hand, autocorrelation in trade signs induces persistence in the price processes, while on the other hand, the liquidity stored in the order book induces mean-reversion. These two effects counterbalance each other exactly. This subtle equilibrium between liquidity takers and liquidity providers, which guarantees price diffusivity at short lags, is not accounted for by the simple Poisson order book model that is simulated here, which explains our observations of anomalous diffusions at short time scales (see also Smith et al. 2003). Because of the absence of positive autocorrelation in trade signs in the model, this effect is magnified when one looks at trades. The next subsection elaborates on this point.

### 9.2.4 Anomalous diffusion at short time scales

We propose a heuristic argument for the understanding of the discrepancy between the model and the data signature plots at short time scales. In what follows, we use the trade time, i.e. the $t$-th trade occurs at time $t$. Denote by $P^{Tr}(t)$ the price of the trade at time $t$, and $\alpha(t)$ its sign:

$$\alpha(t) = \begin{cases} 1 & \text{for a buyer initiated trade, i.e. a buy market order,} \\ -1 & \text{for a seller initiated trade, i.e. a sell market order.} \end{cases}$$

We assume that the two signs are equally probable (symmetric model). But to make the argument valid for both the model (for which successive trade signs are independent) and the data (for which trade signs exhibit long memory) we do not assume independence of successive trade signs. Let $P(t^-)$ and $S(t^-)$ be the mid-price and spread just before the $t$-th trade. Then,

$$P^{Tr}(t) = P(t^-) + \frac{1}{2}\alpha(t)S(t^-). \tag{9.3}$$

For any process $Z$ we define the increment $\Delta Z(t) = Z(t+1) - Z(t)$. With Eq. (9.3), the variance of the trade price process can be written:

$$\begin{aligned}\left(\sigma_1^{Tr}\right)^2 &= \mathbf{V}[\Delta P^{Tr}(t)] \\ &= \mathbf{E}\left[\left(\Delta P^{Tr}(t)\right)^2\right] \\ &= \mathbf{E}\left[(\Delta P(t^-))^2\right] + \mathbf{E}\left[\Delta P(t^-)\Delta(\alpha(t)S(t^-))\right] + \frac{1}{4}\mathbf{E}\left[(\Delta(\alpha(t)S(t^-)))^2\right].\end{aligned}$$

The first term in the right-hand side of the above equation is the variance of mid-price increments, denoted $\sigma_1^2$ thereafter. The second term represents the covariance of mid-price increments and the trade sign weighted by the spread. We may assume that this quantity is negligible. Indeed, this amounts to neglecting the correlation between trade signs and mid-quote movements, which can be justified by the dominance of cancellations and limit orders in comparison to market orders in order book data. We can thus focus on the third term and write:

$$\mathbf{E}\left[(\Delta(\alpha(t)S(t^-)))^2\right] = \mathbf{E}\left[(\alpha(t+1)\Delta S(t^-) + S(t^-)\Delta\alpha(t))^2\right]$$
$$= \mathbf{E}\left[(\Delta\alpha(t))^2\right]\mathbf{E}\left[S(t^-)^2\right] + 2\mathbf{E}\left[\alpha(t+1)\Delta S(t^-)S(t^-)\Delta\alpha(t)\right]$$
$$+ \mathbf{E}\left[\alpha(t+1)^2\right]\mathbf{E}\left[(\Delta S(t^-))^2\right].$$

Again, we neglect the cross term in the right-hand side, which amounts this time to neglect the correlation between trade signs and spread movements. We are thus left with:

$$\mathbf{E}\left[(\Delta(\alpha(t)S(t^-)))^2\right] \approx \mathbf{E}\left[(\Delta\alpha(t))^2\right]\mathbf{E}\left[S(t^-)^2\right] + \mathbf{E}\left[(\Delta S(t^-))^2\right].$$

Finally, if $\rho_1(\alpha)$ is the autocorrelation of trade signs at the first lag, we observe that:

$$\mathbf{E}\left[(\Delta\alpha(t))^2\right] = \mathbf{E}\left[\alpha(t+1)^2\right] + \mathbf{E}\left[\alpha(t)^2\right] - 2\mathbf{E}\left[\alpha(t)\alpha(t+1)\right]$$
$$= 2(1 - \rho_1(\alpha)),$$

and we obtain:

$$\left(\sigma_1^{Tr}\right)^2 \approx \sigma_1^2 + \frac{1}{2}(1 - \rho_1(\alpha))\mathbf{E}\left[S(t^-)^2\right] + \frac{1}{4}\mathbf{E}\left[(\Delta S(t^-))^2\right]. \quad (9.4)$$

More generally, a similar result after $n$ trades may be written:

$$\left(\sigma_n^{Tr}\right)^2 \approx \sigma_n^2 + \frac{1}{2n}(1 - \rho_n(\alpha))\mathbf{E}\left[S(t^-)^2\right]. \quad (9.5)$$

Two effects are clear from Eq. (9.4). First, the trade price variance at short time scales is larger than the mid-price variance. Second, autocorrelation in trade signs dampens this discrepancy. This explains at least partially why the trades signature plot obtained from the data is flatter than the model predictions: $\rho_1(\alpha)_{\text{model}} = 0$, while $\rho_1(\alpha)_{\text{data}} \approx 0.6$. Interestingly, although the arguments that led to (9.4) are rather qualitative, a back of the envelope calculation with $\mathbf{E}\left[S^2\right] \in [1, 9]$ gives a difference $\left(\sigma^{Tr}\right)^2 - \sigma^2$ in the range $[0.5, 4.5]$, which has the same order of magnitude of the values obtained by simulation.

From a modelling perspective, a possible solution to recover the diffusivity, even at very short time scales, is to incorporate long-ranged correlation in the order flow. Tóth et al. (2011) have investigated numerically this route using a "$\epsilon$-intelligence" order book model. In this model, market orders signs are long-ranged correlated, that is, in trade time

$$\rho_n(\alpha) = \mathbf{E}\left[\alpha(t+n)\alpha(t)\right] \propto n^{-\gamma}, \qquad \gamma \in ]0, 1[. \quad (9.6)$$

The size of incoming market orders is a fraction $f$ of the volume displayed at the best opposite quote, with $f$ drawn from the distribution

$$P_\xi(f) = \xi(1-f)^{\xi-1}, \tag{9.7}$$

It is shown in this model that by fine tuning the additional parameters $\gamma$ and $\xi$, one can ensure a diffusive behaviour of the price both at a mesoscopic time scale (a few trades) and a macroscopic time scale (a few hundred trades)[2].

### 9.2.5 Results for CAC 40 stocks

In order to get a cross-sectional view of the performance of the model on all CAC 40 stocks, we estimate the parameters separately for each stock and run a 100,000 event simulation for each parameter set. We then compare in Figure 9.10 the average depth, average spread and the long-term "volatility" measured directly from the data, to those obtained from the simulations. Dashed line is the identity function. It would correspond to a perfect match between model predictions and the data. Solid line is a linear regression $z_{\text{data}} = b_1 + b_2 \, z_{\text{model}}$ for each quantity of interest $z$. Parameters of the regression are given in Table 9.3.

We observe a good agreement between the average depth profiles (Panel $(a)$), and the model successfully predicts the relative magnitudes of the long-term variance $\sigma_\infty^2$ and the average spread $\langle S \rangle$ for different stocks. However, it tends to systematically underestimate $\sigma_\infty^2$ and $\langle S \rangle$. As explained above, this may be related to the absence of autocorrelation in order signs in the model and the presence of more drifting phases in empirical prices than in the simulated ones.

**Table 9.3**  CAC 40 stocks regression results

|  | $b_1$ | $b_2$ | $R^2$ |
|---|---|---|---|
| $\text{Log}\langle A \rangle (5)$ | $-0.42 \, (\pm 0.11)$ | $1.13 \, (\pm 0.04)$ | 0.99 |
| $\text{Log}\langle S \rangle$ | $0.20 \, (\pm 0.06)$ | $1.16 \, (\pm 0.07)$ | 0.97 |
| $\sigma_\infty$ | $-0.012 \, (\pm 0.05)$ | $1.35 \, (\pm 0.11)$ | 0.94 |

## 9.3 Simulation of a Limit Order Book Modelled by Hawkes Processes

The basic order book simulator is now enhanced with arrival times of limit and market orders following mutually exciting Hawkes processes, as in the model described and

---

[2]Note that Toth. el al. Tóth et al. (2011) model the "latent order book", not the actual observable order book. The former represents the *intended* volume at each price level $p$, that is, the volume that would be revealed should the price come close to $p$. So that the interpretation of their parameters, in particular the expected lifetime $\tau_{\text{life}}$ of an order, does not strictly match ours.

mathematically analysed in Chapter 8. We present numerical procedures for the estimation and simulation of Hawkes processes. We show in Section 9.3.4 that using Hawkes process-driven order flows enables a more realistic behaviour of the bid-ask spread than Poissonian order flows.

### 9.3.1 Simulation of the limit order book in a simple Hawkes model

It is known [Large (2007) Da Fonseca and Zaatour (2014b)] that there is a strong clustering of the arrivals of market and limit orders, and we have also seen in Chapter 4 that the flow of limit orders strongly interact with the flow of market order. Such observations naturally advocate for the use of Hawkes processes to model the intensities of submissions of market and limit orders, as was already presented and mathematically studied in Chapter 8.

In this section, we analyze a low-dimensional Hawkes process-based limit order book model. Flows of limit and market orders are represented by two Hawkes processes $N^L$ and $N^M$, with stochastic intensities respectively $\lambda^L$ and $\lambda^M$ defined as:

$$\lambda^M(t) = \lambda_0^M + \int_0^t \alpha_{MM} e^{-\beta_{MM}(t-s)} dN_s^M,$$

$$\lambda^L(t) = \lambda_0^L + \int_0^t \alpha_{LM} e^{-\beta_{LM}(t-s)} dN_s^M + \int_0^t \alpha_{LL} e^{-\beta_{LL}(t-s)} dN_s^L.$$

Three mechanisms can be used here. The first two are self-exciting ones, MM and LL. They are a way to translate into the model the observed clustering of arrival of market and limit orders and the broad distributions of their durations. The third mechanism, LM, is the direct translation of the *market making* property we have identified in Chapter 4. When a market order is submitted, the intensity of the limit order process $N^L$ increases, enforcing the probability that the next event will illustrate a market making behaviour. Note that, for the sake of computational simplicity, we do no implement the reciprocal mutual excitation ML: Although a *market taking* effect has been identified in Chapter 4, it was not observed with all limit orders, but only with the aggressive ones. Since, we preferred to keep the model low-dimensional, the ML effect is not implemented here. Some calibration results based on a more complete model including the market taking effect will be presented in Section 9.4.

### 9.3.2 Algorithm for the simulation of a Hawkes process

We now have to modify our routine for the simulation of a limit order book with Poissonian order flows (Algorithm 1) and replace the simulation of events with exponentially distributed inter-event times (lines 4 and 5 of the algorithm) with the simulation of the Hawkes processes $N^M$ and $N^L$.

Below is a generic algorithm that simulates a $P$-variate Hawkes process with intensities

$$\lambda^n(t) = \lambda_0^n(t) + \sum_{m=1}^{P} \int_0^t \alpha_{nm} e^{-\beta_{nm}(t-s)} dN^m(s), \quad n = 1, \ldots P$$

The simulation is based on a *thinning* method (Lewis and Shedler, 1979). Let $[0, T]$ be the time interval on which the process is to be simulated. We define $I^K(t) = \sum_{n=1}^{K} \lambda^n(t)$ the sum of the intensities of the first $K$ components of the multivariate process. $I^P(t) = \sum_{n=1}^{P} \lambda^n(t)$ is thus the total intensity of the multivariate process and we set $I^0 = 0$.

The detailed routine is given in Algorithm 2.

---

**Algorithm 2** Generic thinning algorithm for the simulation of a multivariate Hawkes process.

---

**Require:** Deterministic base intensities $\lambda^n(t)$ and exponential kernel parameters $(\alpha_{mn})$ and $(\beta_{mn})$, $m, n = 1, \ldots, P$ for the $P$-variate Hawkes process.

1: **Initialization:** Set $i^1 \leftarrow 1, \ldots, i^P \leftarrow 1$ and $I^* \leftarrow I^P(0) = \sum_{n=1}^{P} \lambda_0^n(0)$.
2: **Time of first event:** Draw $s$ exponentially distributed with parameter $I^*$.
3: **while** $s < T$
4: Draw $D$ uniformly distributed on $[0, 1]$.
5: **if** $D \leq \frac{I^P(s)}{I^*}$ **then**
6: Set $t_{i^{n_0}}^{n_0} \leftarrow s$ where $n_0$ is such that $\frac{I^{n_0-1}(s)}{I^*} < D \leq \frac{I^{n_0}(s)}{I^*}$. (New event of type $n_0$)
7: Set $i^{n_0} \leftarrow i^{n_0} + 1$.
8: **end if**
9: **Update maximum intensity:** Set $I^* \leftarrow I^P(s)$. $I^*$ exhibits a jump of size $\sum_{n=1}^{P} \alpha_{nn_0}$ if an event of type $n_0$ has just occurred.
10: **Time of next event:** Draw $s$ exponentially distributed with parameter $I^*$.
11: **end while**

**Ensure:** $\left(\{t_i^n\}_i\right)_{n=1,\ldots,P}$ is a sample path of a multivariate Hawkes process on $[0, T]$.

---

As an illustration we provide some examples of simulations of bivariate Hawkes processes using Algorithm 2. Figure 9.11 shows an example of such a simulation for parameters, and Fig. 9.12 zooms in on a small part of this simulation. Parameters used to compute these graphs are:

$$\lambda_0^1 = 0.1, \alpha_1^{11} = 0.2, \beta_1^{11} = 1.0, \alpha_1^{12} = 0.1, \beta_1^{12} = 1.0,$$
$$\lambda_0^2 = 0.5, \alpha_1^{21} = 0.5, \beta_1^{21} = 1.0, \alpha_1^{22} = 0.1, \beta_1^{22} = 1.0, \quad (9.8)$$

132  Limit Order Books

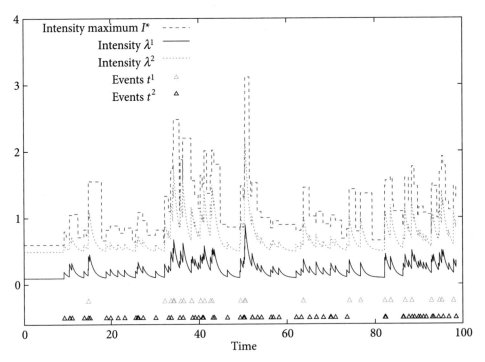

**Fig. 9.11**  Simulation of a two-dimensional Hawkes process with parameters given in Eq. (9.8)

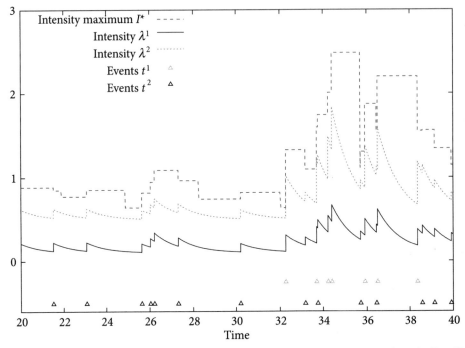

**Fig. 9.12**  Simulation of a two-dimensional Hawkes process parameters given in Eq. (9.8). (Zoom of Fig. 9.11)

### 9.3.3 Parameter estimation

The Hawkes model described above has many variants: By forcing some $\alpha$s to be zero, we can turn off one or several of these features. We therefore have several models to test: namely LM, MM, MM + LM, MM + LL, MM + LL + LM – and try to understand the influence of each effect. As a reference, we will also simulate the model in which $N^M$ and $N^L$ are homogeneous Poisson processes. This variant will be referred to as HP.

Hawkes processes can be estimated by a maximum likelihood method. Details for an efficient computation of the log-likelihood are given in Appendix C.1.1. We fit both $N^L$ and $N^M$ processes by computing on our data these maximum likelihood estimators of the parameters of the different variants of the model. As expected, estimated values varies with the market activity on the day of the sample. However, it appears that estimation of the parameters of stochastic intensity for the MM and LM effect are quite robust. We find an average relaxation parameter $\hat{\beta}_{MM} = 6$, i.e. roughly 170 milliseconds as a characteristic time for the MM effect, and $\hat{\beta}_{LM} = 1.8$, i.e. roughly 550 milliseconds characteristic time for the LM effect. Estimation of models including the LL effect are more troublesome on our data. In the simulations that follows, we assume that the self-exciting parameters are similar ($\alpha_{MM} = \alpha_{LL}$, $\beta_{MM} = \beta_{LL}$) and ensure that the number of market orders and limit orders in the different simulations is roughly equivalent (i.e. approximately 145000 limit orders and 19000 market orders for 24 hours of continuous trading). Table 9.4 summarizes the numerical values used for simulation. Fitted parameters are in agreement with an assumption of asymptotic stationarity. We compute long runs of simulations with our enhanced model, simulating each time 24 hours of continuous trading. With these parameters, the order book is never empty during the simulations. Note however that there is no mechanism to prevent the limit order book from becoming empty. If needed, one can enforce the limits $a_\infty$ and $b_\infty$ for some price far away from the best prices, as in Algorithm 1. Statistics based on the simulation results are discussed in the Section 9.3.4.

**Table 9.4** Estimated values of parameters used for simulations

| Model | $\mu_0$ | $\alpha_{MM}$ | $\beta_{MM}$ | $\lambda_0$ | $\alpha_{LM}$ | $\beta_{LM}$ | $\alpha_{LL}$ | $\beta_{LL}$ |
|---|---|---|---|---|---|---|---|---|
| HP | 0.22 | - | - | 1.69 | - | - | - | - |
| LM | 0.22 | - | - | 0.79 | 5.8 | 1.8 | - | - |
| MM | 0.09 | 1.7 | 6.0 | 1.69 | - | - | - | - |
| MM LL | 0.09 | 1.7 | 6.0 | 0.60 | - | - | 1.7 | 6.0 |
| MM LM | 0.12 | 1.7 | 6.0 | 0.82 | 5.8 | 1.8 | - | - |
| MM LL LM | 0.12 | 1.7 | 5.8 | 0.02 | 5.8 | 1.8 | 1.7 | 6.0 |

| Common parameters: | $m_1^P = 2.7, v_1^P = 2.0, s_1^P = 0.9$ |
|---|---|
| | $V_1 = 275, m_2^V = 380$ |
| | $\lambda^C = 1.35, \delta = 0.015$ |

### 9.3.4 Performances of the simulation

In this section, we present the results of the simulation of the Hawkes process-based model described above. Other than the arrival times of events, there are some differences with the order book simulation described in Algorithm 1: first, the volume distributions $\mathcal{V}^M$ and $\mathcal{V}^L$ are exponential (instead of Log-Gaussian). Second, we do not keep track of the intensities $\lambda_i^L$ for each price level $i$, but use instead one process $N^L$ for the submission of limit order. The submission price of an incoming limit orders is then drawn according to a parametric (Student) distribution centred around the same side best quote and truncated at the opposite best price. Third, the size of these new limit orders is randomly drawn according to an exponential distribution with mean $m_L^V$.

These are minor changes implemented in order to study some alternatives to the choices in Section 9.2, but their influence on the results we present here is clearly moderate, the empahsis being on the arrival times.

With these specifications, we have the following results. Firstly, we can easily check that introducing self- and mutually exciting processes into the order book simulator helps producing more realistic arrival times. Figure 9.13 shows the distributions of the durations of market orders (left) and limit orders (right). As expected, we check that the Poisson assumption has to be discarded, while the use of Hawkes processes helps give more weight to very short time intervals. We also verify that models with only self-exciting processes MM and LL are not able to reproduce the market making feature described in Chapter 4. Distribution of time intervals between a market order and the next limit order are plotted on Fig. 9.14. As expected, no peak for short times is observed if the LM effect is not in the model. However, when the LM effect is included, the simulated distribution of time intervals between a market order and the following limit order is very close to the empirical one.

Besides offering a better simulation of the arrival times of orders, we argue that the LM effect also helps simulating a more realistic behaviour of the bid-ask spread of the order book. On Fig. 9.15, we compare the distributions of the spread for three models – HP, MM, MM+LM – with respect to the empirical measures. We first observe that the model with homogeneous Poisson processes produces a fairly good shape for the spread distribution, but slightly shifted to the right. Small spread values are largely underestimated. When adding the MM effect in order to get a better grasp at market orders' arrival times, it appears that we flatten the spread distribution. One interpretation could be that when the process $N^M$ is excited, markets orders tend to arrive in cluster and to hit the first limits of the order book, widening the spread and thus giving more weight to large spread values. However, since the number of orders is roughly constant in our simulations, there has to be periods of lesser market activity where limit orders reduce the spread. Hence, a flatter distribution. The MM + LM model produces a spread distribution much closer to the empirical shape. It appears from Fig. 9.15 that the LM effect reduces the spread: the market making behaviour helps giving less weight to larger spread values (see the tail of the distribution) and to sharpen the peak of the distribution for small spread values.

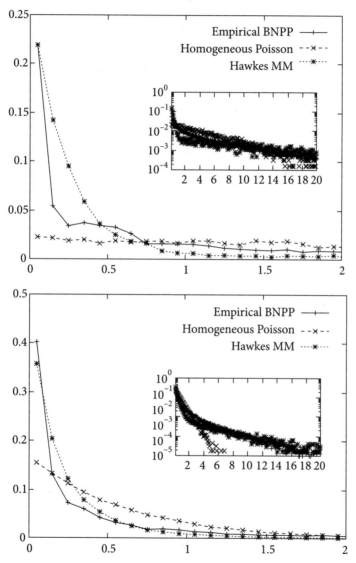

**Fig. 9.13** Empirical density function of the distribution of the durations of market orders (left) and limit orders (right) for three simulations, namely HP, MM, LL, compared to empirical measures. In inset, same data using a semi-log scale. Extracted from Muni Toke (2011)

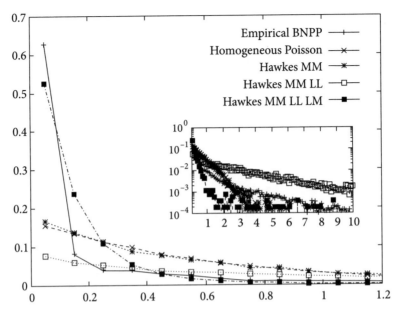

**Fig. 9.14** Empirical density function of the distribution of the time intervals between a market order and the following limit order for three simulations, namely HP, MM+LL, MM+LL+LM, compared to empirical measures. In inset, same data using a semi-log scale. Extracted from Muni Toke (2011)

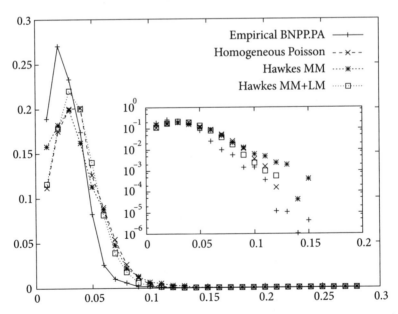

**Fig. 9.15** Empirical density function of the distribution of the bid-ask spread for three simulations, namely HP, MM, MM+LM, compared to empirical measures. In inset, same data using a semi-log scale. X-axis is scaled in euro (1 tick is 0.01 euro). Extracted from Muni Toke (2011)

We show on Fig. 9.16 that the same effect is observed in an even clearer way with the MM + LL and MM + LL + LM models.

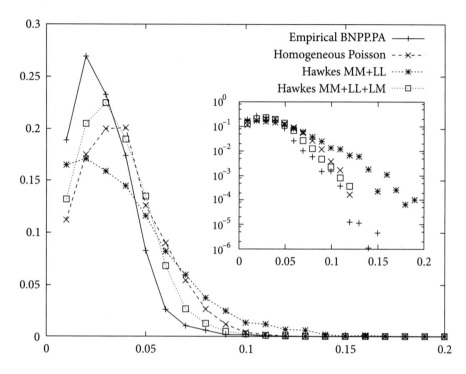

**Fig. 9.16** Empirical density function of the distribution of the bid-ask spread for three simulations, namely HP, MM, MM + LM, compared to empirical measures. In inset, same data using a semi-log scale. X-axis is scaled in euro (1 tick is 0.01 euro). Extracted from Muni Toke (2011)

Actually, the spread distribution produced by the MM + LL model is the flattest one. This is in line with our previous argument. When using two independent self exciting Hawkes processes for arrival of orders, periods of high market orders' intensity gives more weight to large spread values, while periods of high limit orders' intensity gives more weight to small spread values. Adding the cross-term LM to the processes implements a coupling effect that helps reproducing the empirical shape of the spread distribution. The MM + LL + LM simulated spread is the closest to the empirical one.

Finally, it is somewhat remarkable to observe that these variations of the spread distributions are obtained with little or no change in the distributions of the variations of the mid-price. As shown on Fig. 9.17, the distributions of the variations of the mid-price sampled every 30 seconds are nearly identical for all the simulated models.

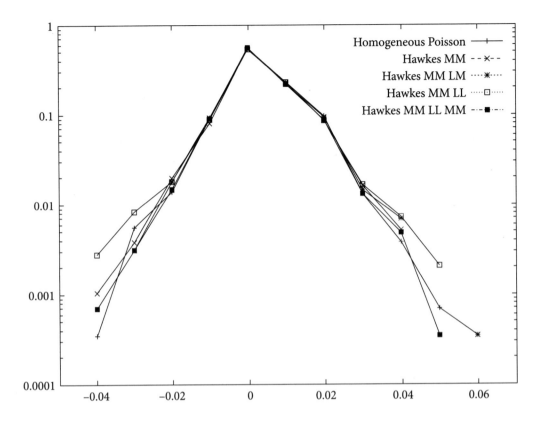

**Fig. 9.17** Empirical density function of the distribution of the 30-second variations of the mid-price for five simulations, namely HP, MM, MM+LM, MM+LL, MM+LL+LM, using a semi-log scale. X-axis is scaled in euro (1 tick is 0.01 euro). Extracted from Muni Toke (2011)

## 9.4 Market Making and Taking, Viewed from a Hawkes-process Perspective

In this short section, we analyze the calibration results of a more general Hawkes process-based model for the limit order book. This time, we distinguish between limit (L) and market (M) orders that change the price (denoted A for aggressive) and those that do not change the price (denoted P for passive); hence we consider four types of orders, denoted by the abbreviations AM, PM, AL and PL. These four types of events are modelled with a four-dimensional Hawkes process $N(t) = (N^{AM}(t), N^{PM}(t), N^{AL}(t), N^{PL}(t))$ with a constant base intensity and an exponential kernel. In other words, the process $N$ has the intensity $\lambda(t) = (\lambda^{AM}(t), \lambda^{PM}(t), \lambda^{AL}(t), \lambda^{PL}(t))$ satisfying:

$$\lambda(t) = \lambda_0 + \int_0^t K(t-u) dN(u), \tag{9.9}$$

where $\lambda_0 = (\lambda_0^{AM}, \lambda_0^{PM}, \lambda_0^{AL}, \lambda_0^{PL})$ is the base intensity and the kernel matrix $K$ has general term $K_{ij}(u) = \alpha_{ij} e^{-\beta_{ij} u}$, $i, j \in \{AM, PM, AL, PL\}$. As before, the model is fitted to the data using a maximum-likelihood estimation described in Appendix C.1.1. In this example, we use 14 days of trading (February 1st to 23rd, 2010) for twelve randomly selected CAC 40 stocks traded on the Paris Bourse. Since, the empirical results in Chapter 4, Section 4.4.1 section has exhibited, as expected, a certain symmetry between the bid and ask sides, we do not distinguish the buy and sell sides and merge all events of the same type from both the bid and ask sides of the book. Following Large (2007), we visualize the results by plotting circles with center coordinates $(\alpha_{ij}, \ln(2)\beta_{ij}^{-1})$ and a diameter proportional to the number of exciting events $j$. Thus, the higher the circle, the stronger the influence of the corresponding event. Similarly, circles on the right side of the graph have a longer influence.

Figure 9.18 plots for the twelve stocks the resulting circles for events that influence the intensity of aggressive limit orders (parameters $\alpha_{AL-j}$ and $\beta_{AL-j}$).

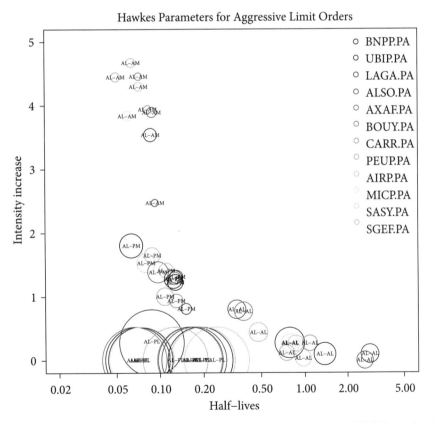

**Fig. 9.18** Hawkes parameters for aggressive limit orders for various CAC40 stocks. These values are computed using MLE estimation on 14 days of trading (Feb.1st-Feb.23rd 2010), 10am-12pm

The intensity of the arrival process of limit orders submitted inside the spread is strongly excited by aggressive market orders, with a rather short half-life. We thus observe here a return of liquidity that tightens the spreads after its widening by an aggressive market order: This is the market making effect already described. Another similar effect of resilience in the order book is observed, a bit less strongly, with passive market orders. The third notable influence is due to the aggressive limit orders, with a less intense effect but longer half-life, illustrating the clustering of these aggressive limit orders. The limited effect of passive limit orders appears in contrast negligible. It is important to remark that this pattern is a general one: All circles of the same type are grouped together on the same part of the graph, i.e. each of the 12 studied stocks exhibit roughly the same behaviour with respect to clustering and market making.

Regarding the reciprocal excitations on aggressive market orders, results are presented on Fig. 9.19.

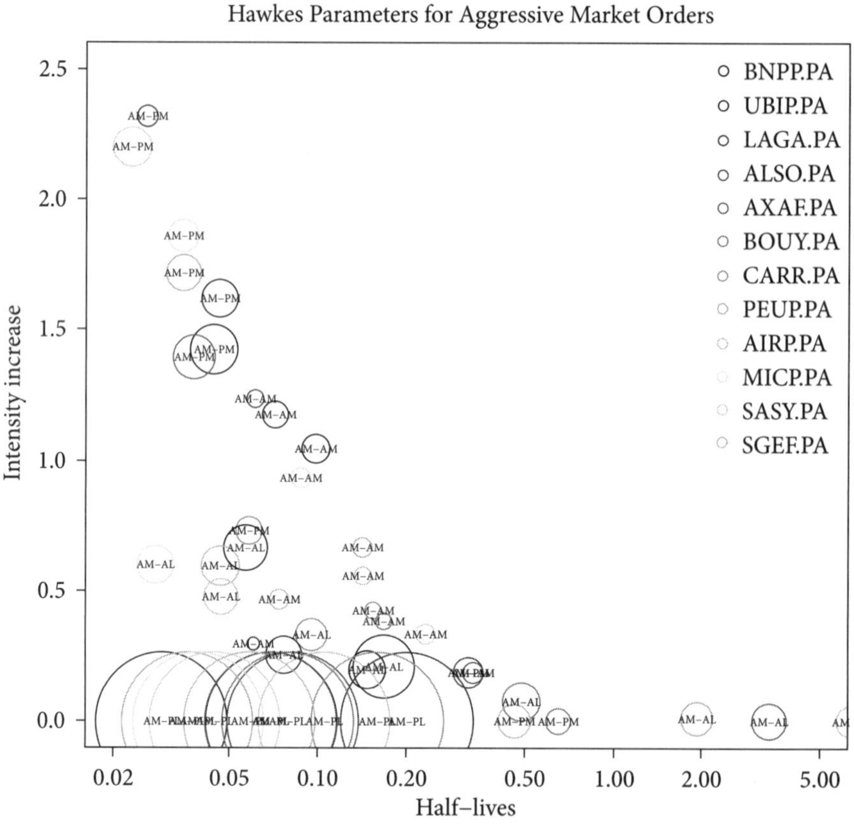

**Fig. 9.19** Hawkes parameters for aggressive market orders for various CAC40 stocks. These values are computed using MLE estimation on 14 days of trading (Feb.1st-Feb.23rd 2010), 10am-12pm

The intensity of the arrival process of market orders that move the price is strongly excited by passive and aggressive market orders. This is an illustration of the clustering of trades, and possibly of a rush to decreasing liquidity: When the volume available at the best limit decreases due to several passive market orders, an aggressive market order is likely to quickly take the remaining liquidity. We also observe a clear influence of aggressive limit orders on aggressive market orders, which corresponds to a market taking effect. This is in line with the observations of Section 4.4.1 based on the use of lagged correlation coefficients. It is however interesting to remark that the strength and length of this effect varies across the stocks studied, i.e. the patterns are less clearly defined for the influence on aggressive market orders than they were in the previous case for the influence of aggressive limit orders.

## 9.5 Conclusion

This chapter was primarily motivated by practical considerations: when using a particular limit order book model, it is important to assess its reliability in reproducing the behaviour of real markets.

Starting with the basic zero-intelligence paradigm and progressing towards more refined models based on Hawkes processes, we have studied limit order book models that can be used to benchmark market making or statistical arbitrage strategies. Note that we do not present results on general state-dependent intensities in this work, and refer the interested reader to some recent contributions such as Huang et al. (2015).

In a different but related direction, a very general, flexible open-source library has been developped by A. Kolotaev in the Chair of Quantitative Finance, and can be found at http://fiquant.mas.ecp.fr. Its purpose is to provide a generic framework for the study of trading strategies in order-driven markets.

# PART FOUR
# IMPERFECTION AND PREDICTABILITY IN ORDER-DRIVEN MARKETS

# CHAPTER 10

# Market Imperfection and Predictability

## 10.1 Introduction

This chapter somewhat departs from our initial motivation of studying limit order books *per se*, and addresses the very practical question of the *predictability* of financial markets based on the information content of limit order books.

Forecasting the market has always been one of the "hottest" topics among market practitioners, and the temptation to identify hopefully profitable signals has never been as high as today. Numerous academic studies aim at identifying some predictive features in the time series of past returns, although many seem to obtain negative results. For instance, it is a well-known stylized fact that there is no evidence of linear correlation between successive returns, see e.g., (Chakraborti et al. 2011a) (Lillo and Farmer, 2004). Such studies seemingly demonstrate the lack of predictive character of the series of past returns, as far as the sign of the next price move is concerned. In that sense, the property generally referred to as the *Efficient Market Hypothesis* does not seem to be challenged.

Rather intriguingly, several books – some popular amongst finance practitioners – introduce and explain predictive strategies that seem to always make money (see e.g., Murphy, 1999; Vidyamurthy, 2004). But, when backtesting those strategies on realistic samples, the results are often quite disappointing, and the strategies no longer profitable. It is likely that the plague of *over-fitting*, inherent to many prediction methods, plays a key role in the seemingly good performances published in those books.

However, there exist several ways to actually make better predictions than just using the series of past returns. For instance, Abergel and Politi (2013) exhibit some synthetic baskets that are not traded and therefore, not necessarily arbitrage-free. Based on these baskets, they provide evidence of short-term predictability. More specific to the context of order-driven markets, the use of limit order book data has yielded interesting prediction results (Zheng et al. 2012; Anane et al. 2015; Anane and Abergel, 2015; Cont et al. 2014).

The study presented in this chapter is performed both from an academic and a professional perspective. It is based on an extensive use of market data, inclusive of limit order book data, and aims at identifying signals that can be used as forecasting tools, and studying their performances. Several prediction methods are introduced and systematically benchmarked. For each prediction method, the statistical properties of the corresponding signals are briefly investigated and the performances of some associated investment strategies are presented.

## 10.2 Objectives, Methodology and Performances Measures

### 10.2.1 Objectives

We focus on the EUROSTOXX 50 European liquid stocks. One year (2013) of full daily order book data are used to achieve the study. For a stock with a mid price $S_t$ at time $t$, the return to be predicted over a period $\delta t$ is $\ln\left(\frac{S_{t+\delta t}}{S_t}\right)$. At time $t$, one can use all the available data for any time $s \leq t$ to perform the prediction.

The focus is on predicting the stocks' returns over a fixed period $\delta t$ using some limit order book indicators. Once the returns and the indicators are computed, the data are sampled on a fixed time grid from 10:00 a.m. to 5:00 p.m. with a resolution $\delta t$. Three different resolutions are tested: 1, 5 and 30 minutes. Below are the definitions of the studied indicators and the rationale behind using them to predict the returns.

**Past return** The past return is defined as $\ln\left(\frac{S_t}{S_{t-\delta t}}\right)$. Two effects justify the use of the past return indicator to predict the next return: The *mean-reversion* effect and the *momentum* effect. If a stock suddenly shows an abnormal return that makes the stock price significantly deviate from its historical mean value, then the mean reversion effect is observed when another large return with opposite sign occurs rapidly after, driving the stock price back to its usual average range. On the other hand, if the stock exhibits, in a progressive fashion, a significant deviation, then the momentum effect occurs when more and more market participants become convinced of the relevance of the move and trade in the same sense, thereby increasing the deviation.

**Order book imbalance** A weighted measure of liquidity on the bid (respectively ask) side is defined as $Liq_{bid} = \sum_{i=1}^{5} w_i |b_i| P_i^B$ (respectively $Liq_{ask} = \sum_{i=1}^{5} w_i a_i P_i^A$), where $P_i^B$ (respectively $P_i^A$) is the price at the limit $i$ on the bid (respectively ask) side, the $a_i$'s and $b_i$'s are the signed quantities, and $w_i$ is a positive, decreasing function of $i$. The maximum number of limits used in the computation (here, 5) reflects the number of visible limits on the trader's screen[1]. Those indicators measure the volume instantaneously available for

---
[1] Note that only non-empty limits are used in this indicator, so that we slightly depart from the notations introduced in Chapter 6, where the index $i$ measured the distance in ticks from the best opposite quote

trading on each side of the order book. Finally, the order book imbalance is defined as $\ln\left(\frac{Liq_{bid}}{Liq_{ask}}\right)$. This indicator summarizes the order book static state and gives an idea about the buy-sell instantaneous equilibrium. When this indicator is significantly higher (respectively lower) than 0, the available quantity at the bid side is significantly higher (respectively lower) than the one at the ask side; only few participants are willing to sell (respectively buy) the stock, which might reflect a market consensus that the stock will move up (respectively down).

**Flow quantity** This indicator summarizes the order book dynamic over the last period $\delta t$. $Q_b$ (respectively $Q_s$) is denoted as the sum of the bought (respectively sold) quantities, over the last period $\delta t$ and the flow quantity is defined as $\ln\left(\frac{Q_b}{Q_s}\right)$. This indicator is similar to the order flow and shows a high positive autocorrelation. The rationale behind using the flow quantity is to verify if the persistence of the flow is informative about the next return.

**EMA** For a process $(X)_{t_i}$ observed at discrete times $(t_i)$, the exponential moving average $EMA(d, X)$ with delay $d$ is defined as $EMA(d, X)_{t_0} = X_{t_0}$ and, for $i \geq 1$, $EMA(d, X)_{t_i} = \omega X_{t_i} + (1 - \omega) EMA(d, X)_{t_{i-1}}$, where $\omega = min(1, \frac{t_i - t_{i-1}}{d})$. The EMA is a weighted average of the process with an exponential decay. The smaller $d$ is, the shorter the $EMA$ memory is.

### 10.2.2 Methodology

We empirically test the market efficiency by predicting the stocks' returns over three different time intervals: 1, 5 and 30 minutes. In Section 10.3, the indicators are either the past returns, the order book imbalance or the flow quantity. A simple method based on historical conditional probabilities is used to assess, separately, the informative effect of each indicator. In Section 10.4, the three indicators and their $EMA(X, d)$ for $d \in \{2^i : i = 0, \ldots, 8\}$ are combined in order to perform a better prediction than that based on a single indicator. Different methods, based on linear regression, are tested. In particular, some statistical and numerical stability problems of the linear regression are addressed.

The predictions are tested statistically, then used to design a simple trading strategy. The goal is to verify whether one can find a profitable strategy covering trading costs of 0.5 basis point[2]. This trading cost is realistic and corresponds to many funds, brokers, and banks trading costs. The possibility of determining, if it exists, a strategy that stays profitable after paying the costs, provides an empirical counter-example to market efficiency. Notice that, in all the sections, the learning samples are sliding windows containing sufficient number of days, and the testing samples are the next days. The models parameters are fitted *in-sample* on a learning sample, and the strategies are tested *out-of-sample* on a testing sample. The sliding training windows prevents the methodology from any over-fitting, since, performances are only computed out of sample.

---

[2]Recall that a *basis point* is a equal to $10^{-4}$ times the current asset price

## 10.2.3 Performance measures

In most studies addressing market efficiency, results are summarized in a linear correlation coefficient. However, such a measure is not sufficient to conclude about returns predictability or market efficiency: Any interpretation of the results should depend on the predicted signal and a corresponding trading strategy. From now on, we shall adopt the very empirical, but quite realistic, view that returns are considered predictable - and thus, the market is considered inefficient - if one can run a profitable strategy covering the trading costs.

## 10.3 Conditional Probability Matrices

Let $Y$ be the variable we want to predict, and $X$ the explanatory variable (or indicator). The conditional probability matrices provide empirical estimates of the conditional probability distribution of $Y$ given $X$. To apply this method, the data need to be discretized in a small number of classes. Let $\{C_i^X : i = 1, \ldots, S_X\}$ be the partition of the state space of $X$ in $S_X$ classes, and $\{C_j^Y : j = 1, \ldots, S_Y\}$ the partition of the state space of $Y$ in $S_Y$ classes. For a given learning period $[0, T]$ containing $N$ observations, let $t_n, n = 1, \ldots, N$ be the time of the $n$-th observation, and $(X_{t_n}, Y_{t_n})$ be the $n$-th observed value of $(X, Y)$. The matrix $M_T$ of occurrences of events up to time $T$ has coefficients $M_T(i, j)$, $1 \leq i \leq S_X$, $1 \leq j \leq S_Y$, defined as:

$$M_T(i, j) = \mathrm{card}\left(\{n : n \leq N, X_{t_n} \in C_i^X, Y_{t_n} \in C_j^Y)\}\right).$$

Then, a prediction of the "next" return $Y_T$ conditional to the observations $X_T$ at time $T$ can be computed using the matrix $M_T$.

For example, to check if the past returns $X$ can help predicting the future returns $Y$, the returns are classified into two classes and the empirical occurrences matrix is computed. Table 10.1 shows the results for the 1-minute returns of Deutsche Telekom over the year 2013.

**Table 10.1** Historical occurrences matrix for Deutsche Telekom over 2013

|       | $Y < 0$ | $Y > 0$ |
|-------|---------|---------|
| $X < 0$ | 19,950 | 21,597 |
| $X > 0$ | 21,597 | 20,448 |

**Table 10.2** Monthly historical conditional probabilities: In the most cases, $\mathbf{P}(Y < 0|X < 0)$ and $\mathbf{P}(Y > 0|X > 0)$ are lower than 50% where $\mathbf{P}(Y > 0|X < 0)$ and $\mathbf{P}(Y < 0|X > 0)$ are higher than 50%

| Month | Jan | Feb | Mar | Apr | May | Jun |
|---|---|---|---|---|---|---|
| $\mathbf{P}(Y < 0\|X < 0)$ | 0.49 | 0.48 | 0.47 | 0.48 | 0.47 | 0.50 |
| $\mathbf{P}(Y > 0\|X < 0)$ | 0.51 | 0.52 | 0.53 | 0.52 | 0.53 | 0.50 |
| $\mathbf{P}(Y < 0\|X > 0)$ | 0.50 | 0.52 | 0.51 | 0.53 | 0.52 | 0.49 |
| $\mathbf{P}(Y > 0\|X > 0)$ | 0.50 | 0.48 | 0.49 | 0.47 | 0.48 | 0.51 |
| Month | Jul | Aug | Sep | Oct | Nov | Dec |
| $\mathbf{P}(Y < 0\|X < 0)$ | 0.48 | 0.51 | 0.51 | 0.47 | 0.46 | 0.44 |
| $\mathbf{P}(Y > 0\|X < 0)$ | 0.52 | 0.49 | 0.49 | 0.53 | 0.54 | 0.56 |
| $\mathbf{P}(Y < 0\|X > 0)$ | 0.51 | 0.50 | 0.51 | 0.50 | 0.51 | 0.55 |
| $\mathbf{P}(Y > 0\|X > 0)$ | 0.49 | 0.50 | 0.49 | 0.50 | 0.49 | 0.45 |

The historical probability to observe a negative return is $\mathbf{P}(X < 0) = 49.70\%$ and to observe a positive return is $\mathbf{P}(X > 0) = 50.30\%$. Therefore, a trader always buying the stock would have a success rate of 50.30%. Also note that $\mathbf{P}(Y < 0|X < 0) = 48.02\%$, $\mathbf{P}(Y > 0|X < 0) = 51.98\%$, $\mathbf{P}(Y < 0|X > 0) = 51.37\%$, $\mathbf{P}(Y > 0|X > 0) = 48.63\%$. Thus, a trader playing the mean-reversion (buy when the past return is negative and sell when the past return is positive), would have a success rate of 51.67%. The same approach, when trading the strategy over 500 stocks, gives a success rate of 54.38% for the buy strategy and of 72.91% for the mean reversion strategy. This simple test shows that the smallest statistical bias can be profitable and useful for designing a trading strategy. However the previous strategy is not realistic: the conditional probabilities are computed in-sample and the full data set of Deutsche Telekom was used for the computation. In reality, predictions have to be computed using only the past data. It is, thus, important to have stationary probabilities. Table 10.2 shows that the monthly observed frequencies are quite stable, and thus can be used to estimate out-of-sample probabilities. Each month, one can use the observed frequencies of the previous month as an estimator of current month probabilities. In the following paragraphs, frequencies matrices are computed on sliding windows for the different indicators. Several classification and prediction methods are presented.

### 10.3.1 Binary case

In the binary case, explanatory variables $X$ are classified in $S_X = 2$ classes, relatively to their historical mean $\overline{X}$: $C_1^X = ]-\infty, \overline{X}]$, $C_2^X = ]\overline{X}, +\infty[$. Using the frequency (occurrences) matrix, a predictor $\widehat{Y}$ of the variable $Y$ is computed as:

$$\widehat{Y} = \begin{cases} \mathbf{E}\left[Y | X \in C_1^X\right] & \text{if } X_T \in C_1^X \\ \mathbf{E}\left[Y | X \in C_2^X\right] & \text{if } X_T \in C_2^X \end{cases}. \tag{10.1}$$

In what follows, we present the results for the prediction of the log-returns $Y$ using for the explanatory variable $X$ one of the quantities defined in Section 10.2: the past returns, the order book imbalance, or the flow quantity. The quality of the prediction is evaluated using four different criteria:

- **AUC** (Area under the curve) combines the true positive rate and the false positive rate to give an idea about the classification quality;
- **Accuracy** is defined as the ratio of the correct predictions ($Y$ and $\widehat{Y}$ have the same sign);
- **Gain** is computed on a simple strategy to measure the prediction performance. Predictions are used to run a strategy that buys when the predicted return is positive and sells when it is negative. At each time, for each stock the strategy's position is in $\{-100,000, 0, +100,000\}$;
- **Profitability** is defined as the gain divided by the traded notional of the strategy presented above. This measure is useful to estimate the gain with different transaction costs.

Figure 10.1 summarizes the results obtained when predicting the 1-minute returns using the three indicators. For each predictor, the AUC and the accuracy are computed over all the stocks. Results are computed over more than 100,000 observations and the amplitude of the 95% confidence interval is around 0.6%. For the three indicators, the accuracy and the AUC are significantly higher than the 50% random guessing threshold. The graph also shows that the order book imbalance gives the best results, and that the past returns is the least successful predictor. Detailed results per stock are given in Appendix D.

In Fig. 10.2, the performances of the trading strategies based on the prediction of the 1-minute returns are presented. The strategies are profitable and the results confirm the predictability of the returns (see the details in Appendix D).

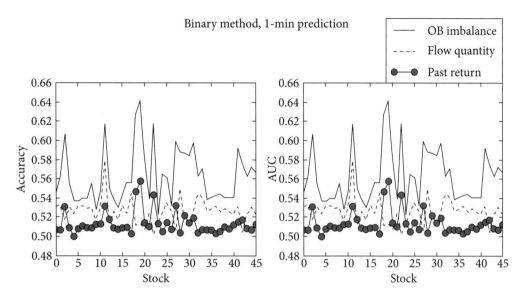

**Fig. 10.1** The quality of the binary prediction: The AUC and the Accuracy are higher than 50%. The three predictors are better than random guessing and are significantly informative. Extracted from Anane and Abergel (2015)

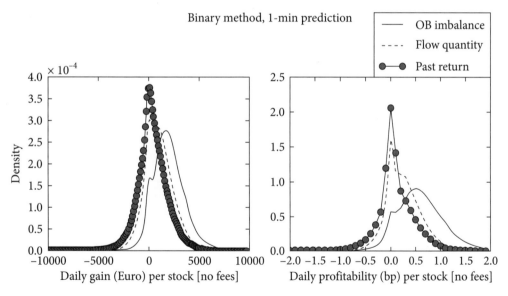

**Fig. 10.2** The quality of the binary prediction: For the 3 predictors, the densities of the gain and the profitability are positively biased, confirming the predictability of the returns. Extracted from Anane and Abergel (2015)

In Fig. 10.3, the cumulative gains of the strategies based on the three indicators over the whole year 2013 are represented. When trading without costs, predicting the 1-minute return using the past return and betting 100,000 euros at each time, would make a 5-million

Euro profit. Even better, predicting using the order book imbalance would make more than 20 million Euros profit. The results confirm the predictability of the returns, but not the inefficiency of the market. In fact, Fig. 10.4 shows that, when adding the 0.5 bp trading costs, only the strategy based on the order book imbalance remains (marginally) positive. Thus, no conclusion, about the market efficiency, can be made (see more details in Appendix D).

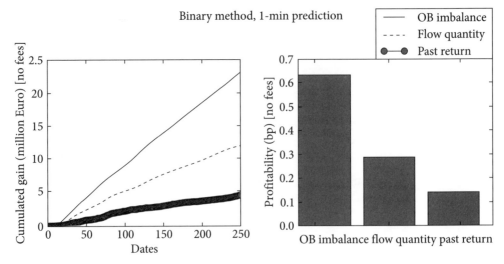

**Fig. 10.3** The quality of the binary prediction: The graphs confirm that the 3 indicators are informative and that the order book imbalance indicator is the most profitable. Extracted from Anane and Abergel (2015)

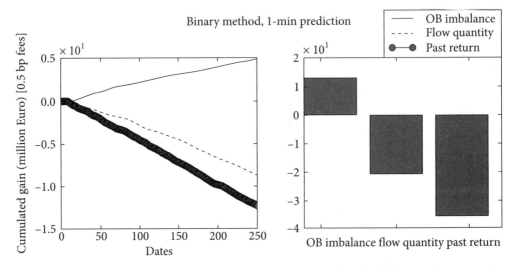

**Fig. 10.4** The quality of the binary prediction: When adding the 0.5 bp trading costs, the strategies are only slightly profitable. Extracted from Anane and Abergel (2015)

Figure 10.5 represents the cumulative gain and the profitability for the 5-minute and the 30-minute strategies (with the trading costs). The strategies are not profitable. Moreover, the predictive power decreases as the time horizon increases.

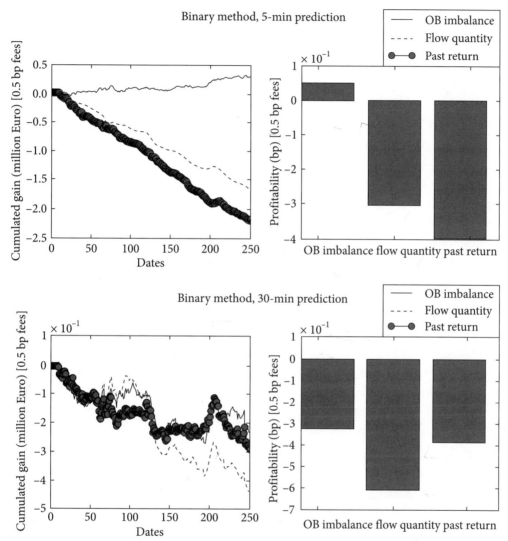

**Fig. 10.5** The quality of the binary prediction: The strategies are not profitable. Moreover, the performances decreases significantly compared to the 1-minute horizon. Extracted from Anane and Abergel (2015)

The results of the binary method show that the returns are significantly predictable. Nevertheless, the strategies based on those predictions are not sufficiently profitable to cover the trading costs. In order to enhance the predictions, the same idea is applied to the four-class case. Moreover, a new strategy based on a minimum threshold of the expected return is tested.

## 10.3.2 Four-class case

We now investigate the case where the explanatory variable $X$ is classified into four classes; "very low values" $C_1^X$, "low values" $C_2^X$, "high values" $C_3^X$ and "very high values" $C_4^X$. As in the binary case, at each time $t_n$, $Y$ is predicted as $\widehat{Y} = E(Y|X \in C_i^X)$, where $C_i^X$ is the class of the current observation $X_{t_n}$. The expectation is estimated from the historical frequencies matrix.

In this four-class case however, a new trading strategy is tested. The strategy is to buy (respectively sell) 100,000 Euros when $\widehat{Y}$ is positive (respectively negative) and $|\widehat{Y}| > \theta$, where $\theta$ is a minimum threshold (we will use $\theta = 1$ basis point in what follows). Notice that the case $\theta = 0$ corresponds to the strategy tested in the binary case.

The rationale for choosing $\theta > 0$ is clearly to avoid trading the stock when the signal is noisy. In particular, when analysing the expectations of $Y$ relative to the different classes of $X$, it is always observed that the absolute value of the expectation is high when $X$ is in one of its extreme classes ($C_1^X$ or $C_4^X$). On the other hand, when X is in one of the intermediary classes ($C_2^X$ or $C_3^X$) the expectation of $Y$ is close to 0 reflecting a noisy signal.

For each indicator $X$, the classes are defined as $C_1^X = ]-\infty, X_a]$, $C_2^X = ]X_a, X_b]$, $C_3^X = ]X_b, X_c]$ and $C_4^X = ]X_c, +\infty[$. To compute $X_a, X_b$ and $X_c$, the 3 following classifications were tested:

- **Quartile classification** The quartile $Q_1$, $Q_2$ and $Q_3$ are computed in-sample for each day, then averaged over the days. $X_a, X_b$ and $X_c$ corresponds, respectively, to $\overline{Q_1}, \overline{Q_2}$ and $\overline{Q_3}$;

- **K-means classification** The K-means algorithm (Hastie et al. 2011), applied to the in-sample data with $k = 4$, gives the centres $G_1, G_2, G_3$ and $G_4$ of the optimal (in the sense of the minimum within-cluster sum of squares) clusters. $X_a, X_b$ and $X_c$ are given respectively by $\frac{G_1+G_2}{2}, \frac{G_2+G_3}{2}$ and $\frac{G_3+G_4}{2}$;

- **Mean-variance classification** The average $\overline{X}$ and the standard deviation $\sigma(X)$ are computed in the learning period. Then, $X_a, X_b$ and $X_c$ correspond, respectively, to $\overline{X} - \sigma(X), \overline{X}$ and $\overline{X} + \sigma(X)$.

Only the results based on the mean-variance classification are presented here, since, the results computed using the two other classifications are equivalent and the differences do not affect the conclusions.

Figure 10.6 compares the profitabilities of the binary and the 4-class methods. For the 1-minute prediction, the results of the 4-class method are significantly better. For the longer horizons, the results of both methods are equivalent. Notice also that, using the best indicator, in the 4-class case, one obtains a significant profit after paying the trading costs. Some more detailed results are given in Appendix D.

The interesting result of this first section is that even when using the simplest statistical learning method, the used indicators are informative and provide a better prediction than random guessing. However, in most cases, the obtained performances are too low to conclude about the market inefficiency.

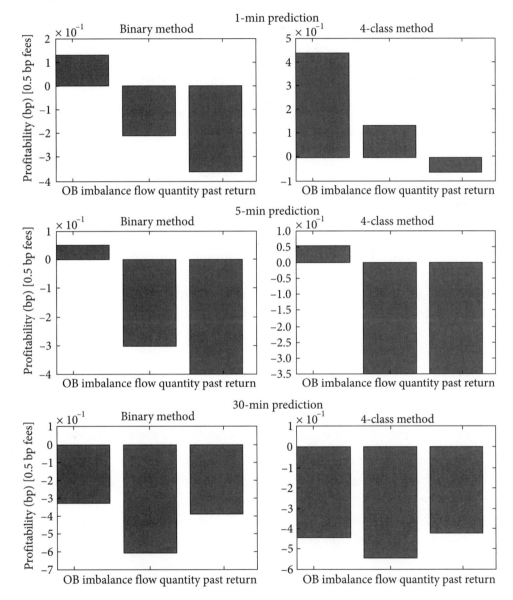

**Fig. 10.6** The quality of the 4-class prediction: For the 1-minute prediction, the results of the 4-class method are significantly better than the results of the binary one. For longer horizons, both strategies are not profitable when adding the trading costs. Extracted from Anane and Abergel (2015)

In order to enhance the performances, the three indicators and their exponential moving averages are combined in the next section.

## 10.4 Linear Regression

In this section, $X$ denotes a 30-column matrix containing the 3 indicators and their $EMA(d)$ for $d \in \{2^i : i = 0, \ldots, 8\}$, and $Y$ denotes the target vector to be predicted. The general approach is to calibrate, on the learning sample, a function $f$ such that $f(X)$ is "the closest possible" to $Y$, and hope that, after the learning period, the relation between $X$ and $Y$ is still well described by $f$. Hence, $f(X)$ would be a good estimator of $Y$. In the linear case, $f$ is supposed to be a linear function and the model errors are supposed to be independent and identically distributed (Seber and Lee, 2003). Actually, the standard textbook model posits a relationship of the form $Y = X\beta + \epsilon$ where $\epsilon$ is Gaussian with mean 0 and variance $\sigma^2$. For numerical reasons, the computations in what follows will be done with z-scored (i.e. scaled and centered) data $\frac{X_i - \overline{X_i}}{\sigma(X_i)}$ instead of $X_i$.

### 10.4.1 Ordinary least squares (OLS)

The OLS method consists in estimating the unknown parameter $\beta$ by minimizing a cost function $J_\beta$ equal to the sum of squares of the residuals between the observed variable $Y$ and the linear approximation $X\beta$. With the usual notation $\|\cdot\|_2$ for the $l^2$-norm, we have $J_\beta = \|Y - X\beta\|_2^2$, and the estimator $\widehat{\beta}$ is thus defined as

$$\widehat{\beta} = \arg\min_{\beta} \left( \|Y - X\beta\|_2^2 \right).$$

This criterion is reasonable if at each time $i$ the row $X_i$ of the matrix $X$ and the observation $Y_i$ of the vector $Y$ represent independent random sample from their populations. The cost function $J_\beta$ depends quadratically on $\beta$, and the critical point equation yields the unique solution

$$\widehat{\beta} = ({}^tXX)^{-1}{}^tXY,$$

provided that ${}^tXX$ is invertible. The expectation, variance and mean squared error of this estimator can be straightforwardly computed:

$$\mathbf{E}\left[\widehat{\beta}|X\right] = \beta,$$
$$\mathbf{Var}\left[\widehat{\beta}|X\right] = \sigma^2({}^tXX)^{-1},$$
$$\mathbf{MSE}\left[\widehat{\beta}\right] = \mathbf{E}\left[\|\widehat{\beta} - \beta\|_2^2 | X\right] = \sigma^2 \sum_i \lambda_i^{-1},$$

where the $\lambda_i$'s are the eigenvalues of $^tXX$. Notice that the OLS estimator is unbiased, but can exhibit an arbitrary high MSE when the matrix $^tXX$ has small eigenvalues.

In the out-of-sample period, $\widehat{Y} = X\widehat{\beta}$ is used to predict the target. We resume our case study where the trading strategy is to buy (respectively sell) 100,000 Euros when $\widehat{Y} > 0$ (respectively $\widehat{Y} < 0$). The binary case based on the order book imbalance indicator is taken as a benchmark to measure the quality of the predictions, since, it performed best in the previous section. The linear regression is computed using 30 indicators, including the order book imbalance, thus one may intuitively expect that it will perform at least as well as the binary case. Figure 10.7 compares the profitabilities of the two strategies. The detailed statistics per stock are given in Appendix D.

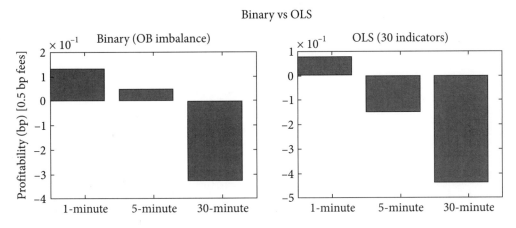

**Fig. 10.7** The quality of the OLS prediction: The results of the OLS method are not better than those of the binary one. Extracted from Anane and Abergel (2015)

Similarly to the binary method, the performance of the OLS method decreases as the horizon increases. But the surprising result is that, when combining all the 30 indicators, the results are not better than just applying the binary method to the order book imbalance indicator. This leads to questioning the quality of the regression.

Figure 10.8 gives some example of the OLS regression coefficients. It is clear that the coefficients are not stable over the time.

For example, for some period, the regression coefficient of the order book imbalance indicator is negative, which does not make any financial sense. It is also observed that, for highly correlated indicators, the regression coefficients might be quite different. This result also does not make sense, since, one would expect to have close coefficients for similar indicators. From a statistical view, this is explained by the high MSE caused by the high colinearity between the variables. In the following paragraphs, this numerical aspect is addressed, and some popular solutions to the OLS estimation problems are tested.

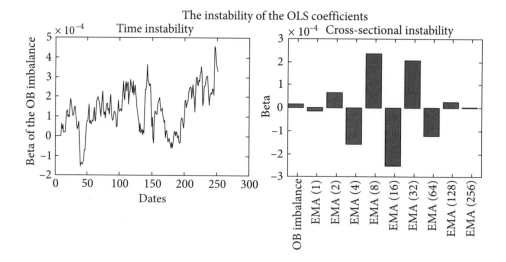

**Fig. 10.8** The quality of the OLS prediction: The graph on the left shows the instability of the regression coefficient of the order book imbalance indicator over the year 2013 for the stock Deutsche Telekom. The graph on the right shows, for a random day, a very different coefficients for similar indicators; the order book imbalance and its exponential moving averages. Extracted from Anane and Abergel (2015)

### 10.4.2 Ridge regression

When solving a linear system $AX = B$, $A$ being invertible, if a small change in the coefficient matrix $(A)$ or a small change in the right hand side $(B)$ causes a large change in the solution vector $(X)$, the system is said to be *ill-conditioned*. An example of an ill-conditioned system is given below:

$$\begin{bmatrix} 1.000 & 2.000 \\ 3.000 & 5.999 \end{bmatrix} \times \begin{bmatrix} x \\ y \end{bmatrix} = \begin{bmatrix} 4.000 \\ 11.999 \end{bmatrix} \Longrightarrow \begin{bmatrix} x \\ y \end{bmatrix} = \begin{bmatrix} 2.000 \\ 1.000 \end{bmatrix}.$$

When making a small change in the matrix $A$:

$$\begin{bmatrix} 1.001 & 2.000 \\ 3.000 & 5.999 \end{bmatrix} \times \begin{bmatrix} x \\ y \end{bmatrix} = \begin{bmatrix} 4.000 \\ 11.999 \end{bmatrix} \Longrightarrow \begin{bmatrix} x \\ y \end{bmatrix} = \begin{bmatrix} -0.400 \\ 2.200 \end{bmatrix}.$$

When making a small change in the vector $B$:

$$\begin{bmatrix} 1.000 & 2.000 \\ 3.000 & 5.999 \end{bmatrix} \times \begin{bmatrix} x \\ y \end{bmatrix} = \begin{bmatrix} 4.001 \\ 11.999 \end{bmatrix} \Longrightarrow \begin{bmatrix} x \\ y \end{bmatrix} = \begin{bmatrix} -3.999 \\ 4.000 \end{bmatrix}.$$

Clearly, it is mandatory to take into consideration such effects before achieving any computation when dealing with experimental data. Various measures of the ill-conditioning of a matrix have been proposed (Riley, 1955), the most popular one probably being (Cheney and Kincaid, 2008) the *condition number* $K(A) = \|A\|_2 \|A^{-1}\|_2$, where $\|\cdot\|_2$ with a matrix argument denotes the induced matrix norm corresponding to the $l^2$ vector norm: $\|A\|_2 = \max_{X \neq 0} \frac{\|AX\|_2}{\|X\|_2}$. The larger $K(A)$, the more ill-conditioned $A$ is. The condition number $K(A)$ gives a measure of the sensitivity of the solution $X$ relative to a perturbation of the matrix $A$ or the vector $B$. More precisely, it is proved that:

- if $AX = B$ and $A(X + \delta X) = B + \delta B$ then $\frac{\|\delta X\|_2}{\|X\|_2} \leq K(A) \frac{\|\delta B\|_2}{\|B\|_2}$ ;

- if $AX = B$ and $(A + \delta A)(X + \delta X) = B$ then $\frac{\|\delta X\|_2}{\|X + \delta X\|_2} \leq K(A) \frac{\|\delta A\|_2}{\|A\|_2}$.

Note that $K(A)$ can be computed as the ratio of the maximum singular value of $A$ over the minimum singular value. Going back to our introductory example above, we have $K(A) = 49988$. The small perturbations can thus be amplified by a factor of almost 50000, causing the instability we have observed.

Figure 10.9 represents the singular values of ${}^tXX$ used to compute the regression of the right graph of Fig. 10.8. The graph shows rapidly decreasing singular values. In particular, the condition number is higher than 80000!

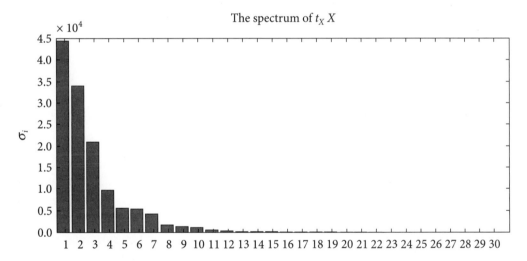

**Fig. 10.9**   The quality of the OLS prediction: The graph shows that the matrix inverted when computing the OLS coefficient is ill-conditioned. Extracted from Anane and Abergel (2015)

This finding explains the instability observed on the previous section. Not only is the performance of the OLS estimator not satisfactory, but the numerical problems caused by the ill-conditioning of the matrix makes the result numerically unreliable. One popular solution to enhance the stability of the estimation of the regression coefficients is the Ridge method. This method was introduced independently by A. Tikhonov, in the context of solving ill-posed problems, around the middle of the 20th century, and by A.E. Hoerl in the context of linear regression. The Ridge regression consists of adding a regularisation term to the original OLS problem:

$$\widehat{\beta_\Gamma} = \arg\min_{\beta} \left( \|Y - X\beta\|_2^2 + \|\Gamma\beta\|_2^2 \right).$$

The new term gives preference to a particular solution with desirable properties. $\Gamma$ is called the Tikhonov matrix and is usually chosen as a multiple of the identity matrix: $\lambda_R I$, where $\lambda_R \geq 0$. The new estimator of the linear regression coefficients is called the Ridge estimator, denoted by $\widehat{\beta_R}$ and defined as follows:

$$\widehat{\beta_R} = \arg\min_{\beta} \left( \|Y - X\beta\|_2^2 + \lambda_R \|\beta\|_2^2 \right).$$

Similarly to the OLS case, straightforward computations show that

$$\widehat{\beta_R} = ({}^tXX + \lambda_R I)^{-1} {}^tXY.$$

Setting $Z = \left(I + \lambda_R({}^tXX)^{-1}\right)^{-1}$ gives $\widehat{\beta_R} = Z\widehat{\beta}$, and we can write after some computations:

$$\mathbf{E}\left[\widehat{\beta_R}|X\right] = Z\beta,$$

$$\mathbf{Var}\left[\widehat{\beta_R}|X\right] = \sigma^2 Z ({}^tXX)^{-1}\, {}^tZ,$$

$$\mathbf{MSE}\left(\widehat{\beta_R}\right) = \mathbf{E}\left[\|(Z\widehat{\beta} - \beta)\|_2^2 | X\right] = \sigma^2 \sum_i \frac{\lambda_i}{(\lambda_i + \lambda_R)^2} + \lambda_R^2\, {}^t\beta({}^tXX + \lambda_R I)^{-2}\beta.$$

The first element of the MSE corresponds exactly to the trace of the covariance matrix of $\widehat{\beta_R}$, i.e. the total variance of the parameters estimations. The second element is the squared distance from $\widehat{\beta_R}$ to $\beta$ and corresponds to the square of the bias introduced when adding the ridge penalty. Note that, when increasing the $\lambda_R$, the bias increases and the variance decreases. On the other hand, when decreasing the $\lambda_R$, the bias decreases and the variance increases, both converging to their OLS values. To enhance the stability of the linear regression, one should compute a $\lambda_R$, such that $\mathbf{MSE}\left(\widehat{\beta_R}\right) \leq \mathbf{MSE}\left(\widehat{\beta}\right)$. As proved in Hoerl and Kennard (1970), this is always possible:

**Theorem** (Hoerl) *There always exist $\lambda_R \geq 0$ such that* $\mathbf{MSE}\left(\widehat{\beta_R}\right) \leq \mathbf{MSE}\left(\widehat{\beta}\right)$.

From a statistical view, adding the Ridge penalty aims at reducing the MSE of the estimator, and is particularly necessary when the covariance matrix is ill-conditioned. From a numerical view, the new matrix to be inverted is ${}^tXX + \lambda_R I$ with as eigenvalues

$(\lambda_i + \lambda_R)_i$. The new condition number satisfies $K({}^tXX + \lambda_R I) = \frac{\lambda_{\max}+\lambda_R}{\lambda_{\min}+\lambda_R} \leq \frac{\lambda_{\max}}{\lambda_{\min}} = K({}^tXX)$. Hence, the ridge regularisation enhances the conditioning of the problem and improves the numerical reliability of the result.

From the previous, it can be seen that increasing the $\lambda_R$ leads to numerical stability and reduces the variance of the estimator, however it increases the bias of the estimator. One has to chose the $\lambda_R$ as a trade-off between those two effects. Next, two estimators of $\lambda_R$ are tested: The Hoerl-Kennard-Baldwin (HKB) estimator Hoerl et al. (1975) and the Lawless–Wang (LW) estimator Lawless and Wang (1976).

In order to compare the stability of the Ridge and the OLS coefficients, Figs 10.10 and 10.11 represent the same test of Fig. 10.8, applied, respectively, to the Ridge HKB and the Ridge LW methods. In the 1-minute prediction case, the graphs show that the Ridge LW method gives the most consistent coefficients. In particular, the coefficient of the order book imbalance is always positive (as expected from a financial point of view) and the coefficients of similar indicators have the same signs.

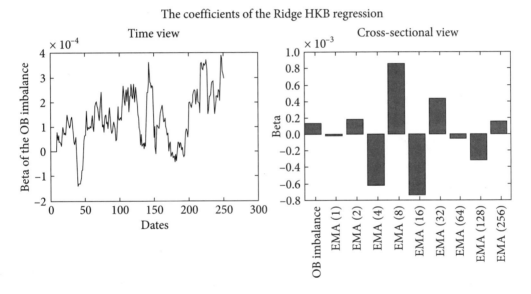

**Fig. 10.10** The quality of the Ridge HKB prediction: The graphs show that the results of the Ridge HKB method are not significantly different from those of the OLS method (Fig. 10.8). In this case, the $\lambda_R$ is close to 0 and the effect of the regularisation is limited. Extracted from Anane and Abergel (2015)

Finally, Fig. 10.12 summarizes the profitabilities of the corresponding strategies of the two methods. Appendix D contains more detailed results per stock.

From the results of this section, it can be concluded that adding a regularisation term to the regression enhances the predictions. The next section deals with an other method of regularisation based on dimension reduction.

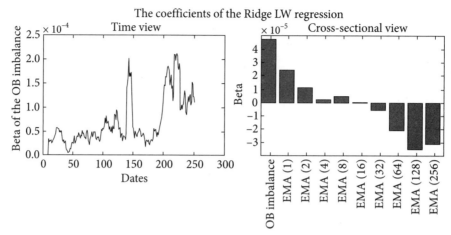

**Fig. 10.11** The quality of the Ridge LW prediction: The graph on the left shows the stability of the regression coefficient of the order book imbalance over the year 2013 for Deutsh Telecom. The coefficient is positive during all the period, in line with the financial view. The graph on the right shows, for a random day, a positive coefficients for the order book imbalance and its short term EMAs. The coefficients decreases with the time; ie the state of the order book "long time ago" has a smaller effect than its current state. More over, for longer than a 10-second horizon, the coefficients become negative confirming the mean-reversion effect. Extracted from Anane and Abergel (2015)

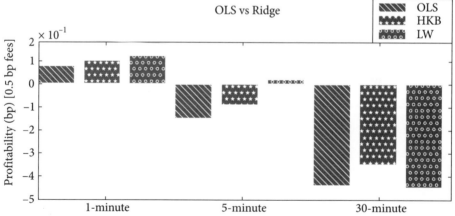

**Fig. 10.12** The quality of the Ridge prediction: For the 1-minute and the 5-minute horizons the LW method performs significantly better than the OLS method. However, for the 30-minute horizon, the HKB method gives the best results. Notice that for the 1-minute case, the LW method improves the performances by 58% compared to the OLS, confirming that stabilizing the regression coefficients (Fig. 10.11 compared to Fig. 10.8), leads to a better trading strategies. Extracted from Anane and Abergel (2015)

## 10.4.3 Least Absolute Shrinkage and Selection Operator (LASSO)

In this paragraph, a simpler, yet very efficient transformation of the original indicators' space, the LASSO regression, is presented. The LASSO method (Tibshirani, 1996) enhances the conditioning of the covariance matrix by reducing the number of the used indicators. Mathematically, the LASSO regression aims to produce a sparse set of regression coefficients – i.e. with some coefficients exactly equal to 0. This is possible thanks to the $l^1$-penalization.

More precisely, the LASSO regression consists in estimating the linear regression coefficient as:

$$\widehat{\beta_L} = \arg\min_{\beta} \left( \|Y - X\beta\|_2^2 + \lambda_L \|\beta\|_1 \right),$$

where $\|\cdot\|_1$ denotes the $l^1$-norm. Writing $|\beta_i| = \beta_{i+} - \beta_{i-}$ and $\beta_i = \beta_{i+} + \beta_{i-}$, with $\beta_{i+} \geq 0$ and $\beta_{i-} \leq 0$, a classic quadratic problem with a linear constraints is obtained and can be solved by a classic solver. We do not have any simple estimator for the parameter $\lambda_L$. We will therefore in this study use a cross-validation method (Hastie et al. 2011) to select the best value of $\lambda_L$ out of the set $\{T 10^{-k} : k \in \{2, 3, 4, 5, 6\}\}$, where $T$ denotes the number of the observations.

Figure 10.13 compares, graphically, the Ridge and the LASSO regularisation, Fig. 10.13 addresses the instability problems observed in Figs 10.8 and 10.15 summarizes the results of the strategies corresponding to the LASSO method. The detailed results per stock are given in Appendix D.

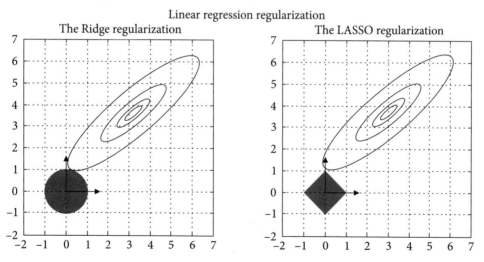

**Fig. 10.13** The quality of the LASSO prediction: The estimation graphs for the Ridge (on the left) and the LASSO regression (on the right). Notice that the $l^1$-norm leads to 0 coefficients on the less important axis. Extracted from Anane and Abergel (2015)

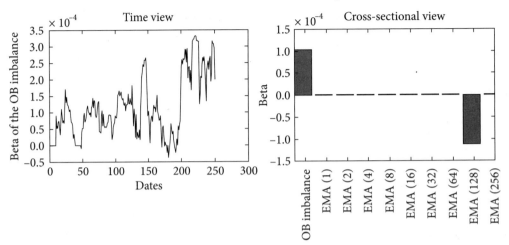

**Fig. 10.14** The quality of the LASSO prediction: The graphs show that the LASSO regression gives a regression coefficients in line with the financial view (similarly to Fig 10.11). Moreover, the coefficients are sparse and simple for the interpretation. Extracted from Anane and Abergel (2015)

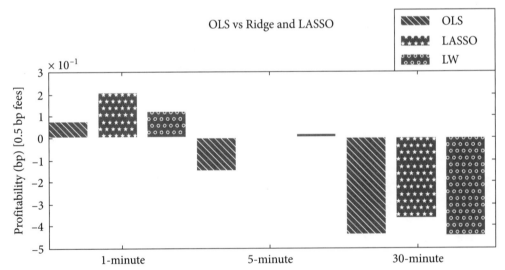

**Fig. 10.15** The quality of the LASSO prediction: Similar as the Ridge regression, the LASSO regression gives a better profitability than the OLS one. Notice that for the 1-minute case, the LASSO method improves the performances by $165\%$ compared to the OLS. Eventhough the LASSO metho is using less regressors than the OLS method, (and thus less signal), the out of sample results are significantly better in the LASSO case. This result confirms the importance of the signal by noise ratio and highlights the importance of the regularisation when addressing an ill-conditioned problem. Extracted from Anane and Abergel (2015)

The next paragraph introduces the natural combination of the Ridge and the LASSO regression and presents this chapter's conclusions concerning the market inefficiency.

### 10.4.4 Elastic net (EN)

The EN regression aims to combine the regularisation effect of the Ridge method and the selection effect of the LASSO one. The idea is to estimate the regression coefficients as:

$$\widehat{\beta}_{EN} = \arg\min_{\beta} \left( \|Y - X\beta\|_2^2 + \lambda_{EN_1} \|\beta\|_1 + \lambda_{EN_2} \|\beta\|_2^2 \right)$$

We will not detail here the details of the estimation of $\widehat{\beta}_{EN}$, which can be found in Zou and Hastie (2005). In this study, the numerical estimation is computed in two steps. In the first step $\lambda_{EN_1}$ and $\lambda_{EN_2}$ are selected via the cross-validation method used in the previous section, and the problem is solved as in the LASSO case. In the second step, the final coefficients are obtained by a Ridge regression ($\lambda_{EN_1} = 0$) over the indicators which had a non-zero coefficient in the first step. The two-step method avoids useless $l^1$-penalty effects on the selected coefficients.

Figure 10.16 shows that the coefficients obtained by the EN method are in line with the financial view and combine both regularisation effects observed when using the Ridge and the LASSO methods.

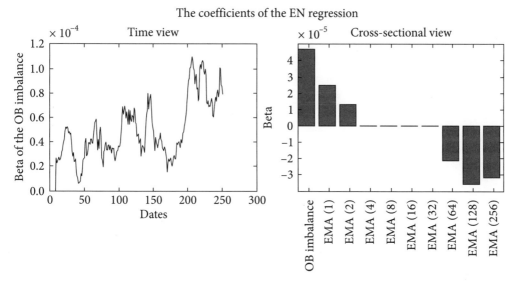

**Fig. 10.16**  The quality of the EN prediction: The graphs show that the EN regression gives a regression coefficients in line with the financial view (similarly to Figs 10.11 and 10.14). Extracted from Anane and Abergel (2015)

Finally, the trading strategy presented in the previous sections (trading only if $\widehat{Y} \geq |\theta|$) is applied to the different regression methods. Figure 10.17 summarizes the obtained results.

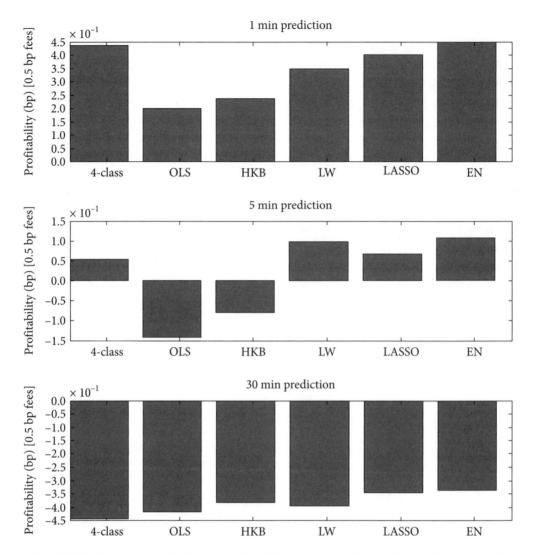

**Fig. 10.17** The quality of the EN prediction: The EN method gives the best results compared to the other regressions. Extracted from Anane and Abergel (2015)

Results for the three time horizons confirm that the predictions of all the regularized method (Ridge, LASSO, EN) are better than the OLS ones. As detailed in the previous paragraphs, this is always the case when the indicators are highly correlated. Moreover, the graphs show that the EN method gives the best results compared to the other regressions. The 1-minute horizon results underline that, when an indicator has an obvious correlation with the target, using a simple method based exhaustively on this indicator, performs as least as well as more sophisticated methods including more indicators. Finally, the performance of the EN method for the 1-minute horizon suggest that the market is inefficient for such horizon. The conclusion is less obvious for the

5-minute horizon. On the other hand, the 30-minute horizon results show that none of the tested methods could find any proof of the market inefficiency for such horizon.

Interpreting these results, one could say that the market is inefficient in the short term, and that this inefficiency progressively disappears as new information becomes more widely spread.

## 10.5 Conclusion

This chapter is a large-scale, empirical study, over the EUROSTOXX 50 universe, testing the predictability of returns. The first part of the study shows that the future returns are not independent of the past dynamic and state of the order book. In particular, the order book imbalance indicator is informative and provides a reliable prediction of the returns. The second part of the study shows that combining different order book indicators using adequate regressions lead to trading strategies with good performances even when paying the trading costs. In particular, our results demonstrate that the market is inefficient in the short term and that a period of a few minutes is necessary for prices to adjust to the new information present in the limit order book.

APPENDIX A

# A Catalogue of Order Types

We list below some examples of orders that exist in different exchanges, along with short descriptions:

- Market order: A market order is an order to buy or sell an asset at the bid or offer price currently available in the marketplace.
- Limit order: A limit order is an order to buy or sell a contract at a specified price or better.
- *Good till date order (GTD)*: An order that remains in the marketplace until it is executed or until the market closes on the date specified.
- *Fill or kill order (FOK)*: An order that must be executed as a complete order immediately, otherwise it is cancelled.
- *Market on close order (MOC)*: A market order submitted to be executed as close to the closing price as possible.
- *Market on open order (MOO)*: A market order to be executed when the market opens.
- *Limit on close order (LOC)*: A limit order to be executed as a market order at the closing price if the closing price is equal to or better than the submitted limit price.
- *Limit on open order (LOO)*: A limit order to be executed as a market order when the the market opens if the opening price is equal to or better than the limit price.
- *Stop order*: An order converted to a market buy or sell order once a specified stop price is attained or penetrate
- *Pegged to market order*: An order that is pegged to buy on the best offer and sell on the best bid.
- *Market to limit order*: an order that is sent in as a market order to be executed at the current best price. If the entire order is not immediately executed at the market price,

the remainder of the order is resubmitted as a limit order with the limit price set to the price at which the original order was executed as a market order.

- *Discretionary order*: An order that allows the broker to delay the execution at her discretion to try and get a better price.
- *Iceberg order*: An order whose (generally large) large volume is only partially disclosed. Iceberg orders belong to the category of "hidden orders", whereby investors wishing to hide large-size orders, can do so by applying the "hidden" attribute to a large volume order and hide the submitted quantity from the market.
- *Block order*: A limit order with a minimum size of 50 contracts.
- *Volume-weighted average price order (VWAP)*: Bid or ask orders to be executed at the volume weighted average price traded in the market of reference for a given security, during a future period of time.

# APPENDIX B

# Limit Order Book Data

An experimental approach to the study of limit order books lies in the availability of data. Most of the results presented in this book - in any case, those we have produced ourselves - use the **Thomson Reuters Tick History** (TRTH) database. All exchange-traded assets worldwide are present in the TRTH database, where they are identified by their **Reuters Identification Code** (RIC). Similar to most historical databases directly provided by the exchanges, the TRTH data come into the form of two separate files, a **trade** file recording all transactions, and an **event** file recording every change in the limit order book. Some very specific information, such as traders' identities, cannot be publicly disclosed for obvious confidentiality reasons, but in theory, one could reconstruct the sequence of order arrivals of all types using this trade and event files.

After explaining the algorithm used for the processing of limit order book data, we describe the specific data sets that have been used at various places in this book. That way, our results can be reproduced, extended and, of course, challenged, based on the very same data sets we have used.

## B.1 Limit Order Book Data Processing

Because one cannot distinguish market orders from cancellations just by observing changes in the limit order book (the "event" file), and since, the timestamps of the "trade" and "event" files are *asynchronous*, we use a matching procedure to reconstruct the order book events.

In a nutshell, we proceed as follows for each stock and each trading day:

1. Parse the "event" file to compute order book state variations:
   - If the variation is positive (volume at one or more price levels has increased), then label the event as a limit order.
   - If the variation is negative (volume at one or more price levels has decreased), then label the event as a "likely market order".

- If no variation—this happens when there is just a renumbering in the field "Level" that does not affect the state of the book—do not count an event.

2. Parse the "trade" file and for each trade:
   - Compare the trade price and volume to likely market orders whose timestamps are in $[t^{Tr} - \Delta t, t^{Tr} + \Delta t]$, where $t^{Tr}$ is the trade timestamp and $\Delta t$ is a predefined, market-dependent time window. For instance, we set $\Delta t = 3$ s for CAC 40 stocks over the year 2011, based on the empirical fact that the median delay in reporting trades is $-900$ ms: half of the trades are reported in the "trade" file 900 milliseconds or less *before* the corresponding change appears in the "event" file.
   - Match the trade to the first likely market order with the same price and volume and label the corresponding event as a market order—making sure the change in order book state happens at the best price limits.
   - Remaining negative variations are labeled as cancellations.

Doing so, we have an average matching rate of around 85% for CAC 40 stocks. As a byproduct, one gets the sign of each matched trade, that is, whether it is buyer or seller-initiated.

Tables B.1 and B.2 below provide an example of the data files and of the matching algorithm

**Table B.1** Tick by tick data file sample. Note that the field "Level" does not necessarily correspond to the distance in ticks from the best opposite quote as there might be gaps in the book. Lines corresponding to the trades in Table B.2 are highlighted in italics

| Timestamp | Side | Level | Price | Quantity |
|---|---|---|---|---|
| 33480.158 | B | 1 | 121.1 | 480 |
| 33480.476 | B | 2 | 121.05 | 1636 |
| 33481.517 | B | 5 | 120.9 | 1318 |
| *33483.218* | *B* | *1* | *121.1* | *420* |
| 33484.254 | B | 1 | 121.1 | 556 |
| 33486.832 | A | 1 | 121.15 | 187 |
| 33489.014 | B | 2 | 121.05 | 1397 |
| *33490.473* | *B* | *1* | *121.1* | *342* |
| *33490.473* | *B* | *1* | *121.1* | *304* |
| *33490.473* | *B* | *1* | *121.1* | *256* |
| 33490.473 | A | 1 | 121.15 | 237 |

**Table B.2** Trades data file sample

| Timestamp | Last | Last quantity |
|---|---|---|
| 33483.097 | 121.1 | 60 |
| 33490.380 | 121.1 | 214 |
| 33490.380 | 121.1 | 38 |
| 33490.380 | 121.1 | 48 |

**Remark B.1** As a comment, we note that, depending on the markets and periods, this matching rate can deteriorate or improve. The possibility of using *hidden* orders is definitely one of the main reasons why the matching rate is not closer to 100%; the increase in trading frequency for the most liquid assets such as equity index futures is another, since the occurrence of trades with the same time stamps increases, while the time resolution of the TRTH database is still the millisecond.

## B.2 Chapter 2

In Chapter 2, we have produced our own empirical plots based on TRTH database for the Paris stock exchange. We have selected four stocks: France Telecom (FTE.PA), BNP Paribas (BNPP.PA), Societe Générale (SOGN.PA) and Renault (RENA.PA). For any given stocks, the data displays time-stamps, traded quantities, traded prices, the first five best-bid limits and the first five best-ask limits. Except when mentioned otherwise, all statistics are computed using all trading days from Oct, 1st 2007 to May, 30th 2008, i.e. 168 trading days. On a given day, orders submitted between 9:05am and 5:20pm are taken into account, i.e. first and last minutes of each trading days are removed.

## B.3 Chapter 3

In Chapter 3, we have used TRTH database for fourteen stocks traded on the Paris stock exchange, from January 4th, 2010 to February 22nd, 2010. The fourteen stocks under investigation are: Air Liquide (AIRP.PA, chemicals), Alstom (ALSO.PA, transport and energy), Axa (AXAF.PA, insurance), BNP Paribas (BNPP.PA, banking), Bouygues (BOUY.PA, construction, telecom and media), Carrefour (CARR.PA, retail distribution), Danone (DANO.PA, milk and cereal products), Lagardére (LAGA.PA, media), Michelin (MICP.PA, tires manufacturing), Peugeot (PEUP.PA, vehicles manufacturing), Renault (RENA.PA, vehicles manufacturing), Sanofi (SASY.PA, healthcare), Vinci (SGEF.PA, construction and engineering), Ubisoft (UBIP.PA, video games). All these stocks except

Ubisoft were included in the CAC 40 French index in January and February 2010, i.e. they are among the largest market capitalizations and most liquid stocks on the Paris stock exchange.

## B.4 Chapter 4

## B.5 Distribution of Durations

In Section 4.3, we use TRTH database for several assets of various types:

- BNP Paribas (RIC: BNPP.PA): 7th component of the CAC40 during the studied period
- Peugeot (RIC: PEUP.PA): 38th component of the CAC40 during the studied period
- Lagardère SCA (RIC: LAGA.PA): 33th component of the CAC40 during the studied period
- Dec.2009 futures on the 3-month Euribor (RIC: FEIZ9)
- Dec.2009 futures on the Footsie index (RIC: FFIZ9)

For each trading day between September 10th, 2009 and September 30th, 2009 (i.e. 15 days of trading), we use 4 hours of data, from 9:30am to 1:30pm. This time frame is convenient for european equity markets because it avoids the opening of American markets and the consequent increase of activity.

In Table B.3, we give the number of market and limit orders detected on our 15 4-hour samples for each studied order book.

**Table B.3** Number of limit and markets orders recorded on 15 samples of four hours (Sep 10th to Sep 30th, 2009; 9:30am to 1:30pm) for 5 different assets (stocks, index futures, bond futures)

| Code | Number of limit orders | Number of market orders |
|---|---|---|
| BNPP.PA | 321,412 | 48,171 |
| PEUP.PA | 228,422 | 23,888 |
| LAGA.PA | 196,539 | 9,834 |
| FEIZ9 | 110,300 | 10,401 |
| FFIZ9 | 799,858 | 51,020 |

On the studied period, market activity ranges from 2.7 trades per minute on the least liquid stock (LAGA.PA) to 14.2 trades per minute on the most traded asset (Footsie futures).

### B.5.1 Lagged correlation matrix

In Section 4.4.1, the data set comprises the 30 constituents of the DAX index traded on the Frankfurt Stock Exchange, and results are computed using four months of tick-by-tick data, from February to June 2014.

## B.6 Chapter 9

The dataset used for the simulations presented in Chapter 9 consists of TRTH database for the CAC 40 index constituents in March 2011 (23 trading days), namely, tick-by-tick order book data up to 10 price levels, and trades. In order to avoid the diurnal seasonality in trading activity (and the impact of the US market open on European stocks), we restrict our attention to the time window $[9:30\text{--}14:00]$ Paris time.

# APPENDIX C

# Some Useful Mathematical Notions

## C.1 Point Processes

Point processes are a class of stochastic processes that appear in a natural fashion when a phenomenon is best described by events occurring at points in time separated by intervals of inactivity. A reference book on the subject is (Daley and Vere-Jones, 2003, 2008). In this brief appendix, we recall some standard notions and notations for point processes.

> **Definition C.1** A *point process* is an increasing sequence $(T_n)_{n \in \mathbb{N}}$ of positive random variables defined on a measurable space $(\Omega, \mathcal{F}, \mathbb{P})$.

We will restrict our attention to processes that are *non-explosive*, that is, for which $\lim_{n \to \infty} T_n = \infty$. To each realization $(T_n)$ corresponds a counting function $(N(t))_{t \in \mathbb{R}^+}$ defined by

$$N(t) = n \text{ if } t \in [T_n, T_{n+1}[, n \geq 0. \tag{C.1}$$

$(N(t))$ is a right continuous step function with jumps of size 1 and carries the same information as the sequence $(T_n)$, so that $(N(t))$ is also called a point process.

> **Definition C.2** A *multivariate point process* (or marked point process) is a point process $(T_n)$ for which a random variable $X_n$ is associated to each $T_n$. The variables $X_n$ take their values in a measurable space $(E, \mathcal{E})$.

We will restrict our attention to the case where $E = \{1, \ldots, M\}$, $m \in \mathbb{N}^*$. For each $m \in \{1, \ldots, M\}$, we can define the counting processes

$$N^m(t) = \sum_{n \geq 1} \mathbb{I}(T_n \leq t) \mathbb{I}(X_n = i). \tag{C.2}$$

We also call the process

$$N(t) = (N^1(t), \ldots, N^M(t))$$

a multivariate point process.

> **Definition C.3** (Intensity of a point process) A point process $(N(t))_{t \in \mathbb{R}^+}$ can be completely characterized by its (conditional) intensity function, $\lambda(t)$, defined as
>
> $$\lambda(t) = \lim_{u \to 0} \frac{\mathbb{P}\left[N(t+u) - N(u) = 1 | \mathcal{F}_t\right]}{u}, \quad (C.3)$$
>
> where $\mathcal{F}_t$ is the history of the process up to time $t$, that is, the specification of all points in $[0, t]$. Intuitively
>
> $$\mathbb{P}\left[N(t+u) - N(u) = 1 | \mathcal{F}_t\right] = \lambda(t)\, u + o(u), \quad (C.4)$$
> $$\mathbb{P}\left[N(t+u) - N(u) = 0 | \mathcal{F}_t\right] = 1 - \lambda(t) u + o(u), \quad (C.5)$$
> $$\mathbb{P}\left[N(t+u) - N(u) > 1 | \mathcal{F}_t\right] = o(u). \quad (C.6)$$
>
> This is naturally extended to the multivariate case by setting for each $m \in \{1, \ldots, M\}$
>
> $$\lambda^m(t) = \lim_{u \to 0} \frac{\mathbb{P}\left[N^m(t+u) - N^m(u) = 1 | \mathcal{F}_t\right]}{u}. \quad (C.7)$$

> **Definition C.4** A point process is stationary when for every $r \in \mathbb{N}^*$ and all bounded Borel subsets $A_1, \ldots, A_r$ of the real line, the joint distribution of
>
> $$\{N(A_1 + t), \ldots, N(A_r + t)\}$$
>
> does not depend on $t$.

### C.1.1 Hawkes processes

The main definitions and fundamental properties of Hawkes processes have been given in Chapter 8 Section C.1.1. Here, we make precise some more specific points: the construction of a Lyapunov function for Markovian Hawkes processes, and the calibration of Hawkes processes based on maximum likelihood estimations.

## Lyapunov functions for Hawkes processes

For the sake of completeness, an explicit construction of a Lyapunov function for a multi-dimensional Hawkes processes $\mathbf{N} = (N^i)$ with intensities

$$\lambda^i(t) = \lambda_0^i + \sum_j \int_0^t \alpha_{ij} e^{-\beta_{ij}(t-s)} dN^j(s)$$

is provided here.

Denote as in Proposition 8.2

$$\mu^{ij}(t) = \int_0^t \alpha_{ij} e^{-\beta_{ij}(t-s)} dN^j(s),$$

so that there holds

$$\lambda^i(t) = \lambda_0^i + \sum_j \mu^{ij}(t). \tag{C.8}$$

We assume the following

$$\forall i, j, \ \alpha_{ij} > 0, \ \beta_{ij} > 0, \tag{C.9}$$

as well as the spectral condition (8.10)

$$\rho(\mathbf{A}) < 1. \tag{C.10}$$

The infinitesimal generator associated to the Markovian process $(\mu^{ij})$, $1 \leqslant i, j \leqslant D$, is the operator

$$\mathcal{L}_H F(\mu) = \sum_j \lambda^j (F(\mu + \Delta^j(\mu)) - F(\mu)) - \sum_{i,j} \beta_{ij} \mu^{ij} \frac{\partial F}{\partial \mu^{ij}},$$

where $\mu$ is the vector with components $\mu^{ij}$ and the $\lambda^j$ are as in (C.8). The notation $\Delta^j(\mu)$ characterizes the jumps in those of the entries in $\mu$ that are affected by a jump of the process $N^j$. For a **fixed** index $j$, it is given by the vector with entries $\alpha_{ij}$ at the relevant spots, and zero entries elsewhere.

A Lyapunov function for the associated semi-group is sought under the form

$$V(\mu) = \sum_{i,j} \delta_{ij} \mu^{ij} \tag{C.11}$$

(since, the intensities are always positive, a linear function will be coercive). Assuming (C.11), there holds

$$\mathcal{L}_H V = \sum_j \lambda^j \left( \sum_i \delta_{ij} \alpha_{ij} \right) - \sum_{i,j} \beta_{ij} \mu^{ij} \delta_{ij}$$

or

$$\mathcal{L}_H V = \sum_{i,j} \left( \lambda_0^j + \sum_k \mu^{jk} \right) \delta_{ij} \alpha_{ij} - \beta_{ij} \mu^{ij} \delta_{ij}. \tag{C.12}$$

At this stage, it is convenient to introduce $\epsilon$ the maximal eigenvector of the matrix $\mathbf{A}$ (introduced in Proposition 8.3) with entries

$$A_{ij} = \frac{\alpha_{ji}}{\beta_{ji}}.$$

Denote by $\kappa$ the associated maximal eigenvalue. By Assumption (8.10), one has that $0 < \kappa < 1$ and furthermore, by Perron-Frobenius theorem, there holds: $\forall i, \epsilon_i > 0$.

Assuming that

$$\delta_{ij} \equiv \frac{\epsilon_i}{\beta_{ij}}, \tag{C.13}$$

the expression for $V$ becomes

$$V(\mu) = \sum_{i,j} \epsilon_i \frac{\mu^{ij}}{\beta_{ij}}. \tag{C.14}$$

Plugging Eq. (C.14) in Eq. (C.12) yields

$$\mathcal{L}_H V = \sum_{i,j} \lambda_0^j \delta_{ij} \alpha_{ij} + \sum_{i,j,k} \mu^{jk} \epsilon_i \frac{\alpha_{ij}}{\beta_{ij}} - \sum_{j,k} \beta_{jk} \mu^{jk} \delta_{jk}$$

$$= \sum_{i,j} \lambda_0^j \delta_{ij} \alpha_{ij} + (\kappa - 1) \sum_{j,k} \epsilon_k \mu^{jk},$$

using the identity $\sum_j \mathbf{A}_{ji} \epsilon_i = \kappa \epsilon_j$. A comparison with Eq. (C.14) easily yields the upper bound

$$\mathcal{L}_H V \leq -\gamma V + C, \tag{C.15}$$

with $\gamma = (1 - K)\beta_{\min}$, $\beta_{\min} \equiv \text{Inf}_{i,j}(\beta_{ij}) > 0$ by assumption, and $C = \sum_{i,j} \lambda_0^j \delta_{ij} \alpha_{ij} \equiv \kappa \epsilon . \lambda_0$.

The following result generalizes the form of Lyapunov functions beyond Eq. (C.11)[1]:

**Lemma C.5** *Under the standing assumptions (8.10) and (C.9), one can construct a Lyapunov function of arbitrary high polynomial growth at infinity.*

**Proof** Let $n \in \mathbb{N}^*$, and $V$ be the function defined in Eq. (C.14). Raising $V$ to the power $n$ yields

$$\mathcal{L}_H(V^n)(\mu) = \sum_j \lambda^j (V^n(\mu + \Delta^j(\mu)) - V^n(\mu)) - nV^{n-1}\left(\sum_{i,j} \beta_{ij}\mu^{ij} \frac{\partial V}{\partial \mu^{ij}}\right). \quad (C.16)$$

Upon factoring $V^n(\mu + \Delta^j(\mu)) - V^n(\mu)$:

$$V^n(\mu + \Delta^j(\mu)) - V^n(\mu) = (V(\mu + \Delta^j(\mu)) - V(\mu))\left(\sum_{k=0}^{n-1} V^{n-1-k}(\mu + \Delta^j(\mu))V^k(\mu)\right),$$

the linearity of $V$ yields the following expression

$$V^n(\mu + \Delta^j(\mu)) - V^n(\mu) = nV^{n-1}(\mu)(V(\mu + \Delta^j(\mu)) - V(\mu)) + M_j(V)(\mu),$$

where $M_j V(\mu)$ can be bounded by a polynomial function of order $n-1$ at infinity in $\mu$. Therefore, one can rewrite (C.16) as follows

$$\mathcal{L}_H(V^n)(\mu) = (nV^{n-1}\mathcal{L}_H(V))(\mu) + M(V)(\mu), \quad (C.17)$$

where $M(V)(\mu)$ is a polynomial of order $n-1$ in $\mu$. Combining Eq. (C.14) with Eq. (C.17) shows that $V^n$ is also a Lyapunov function for the Hawkes process.

**Maximum-likelihood estimation**

We provide some elements for the calibration on market data of Hawkes processes with exponential kernels. Let us consider a sample realization on $[0, T]$ of a $D$-dimensional (generalized) Hawkes process, for which the $m$-th coordinate $N^m$ admits an intensity of the form:

$$\lambda^m(t) = \lambda_0^m(t) + \sum_{n=1}^{D} \int_0^t \sum_{j=1}^{P} \alpha_{mn,j} e^{-\beta_{mn,j}(t-s)} dN^n(s), \quad (C.18)$$

where $\lambda_0^m : \mathbb{R}_+ \to \mathbb{R}_+$ is a deterministic (not necessarily constant) function, the number $P$ of exponential kernels is a fixed integer, and for all $m, n = 1, \ldots, D$, and $j = 1, \ldots, P$, $\alpha_{mn,j}$ and $\beta_{mn,j}$ are positive constants. We will develop the estimation procedure for this process, although in the simpler version of this model used throughout the book, we have

---

[1] See also the construction of an exponentially growing Lyapunov function in Zheng et al. (2014) or Clinet (2015)

set $P = 1$ and $\lambda_0^m(t) = \lambda_0^m$ a positive constant, so that the general defining equation in C.18 reduces to the usual expression:

$$\lambda^m(t) = \lambda_0^m + \sum_{n=1}^{D} \int_0^t \alpha_{mn} e^{-\beta_{mn}(t-s)} dN^n(s). \tag{C.19}$$

(when $P = 1$, $\alpha_{mn,1}, \beta_{mn,1}$ are identical to the $\alpha_{mn}, \beta_{mn}$ previously introduced). Let $\{T_i, Z_i\}_{i=1,\ldots,N}$ be the ordered pool of all $N$ observed events of the sample, where $Z_i \in \{1, \ldots, D\}$ denotes the type of the observed event at time $T_i$.

Let $\{T_i^m\}_{i=1,\ldots,N_m}$ be the extracted ordered sequence of the $N_m$ observed events of type $m$. The log-likelihood $\ln \mathcal{L}$ of the multi-dimensional Hawkes process can be computed as the sum of the likelihood of each coordinate, and is thus written:

$$\ln \mathcal{L}(\{N(t)\}_{t \leq T}) = \sum_{m=1}^{D} \ln \mathcal{L}^m(\{N^m(t)\}_{t \leq T}), \tag{C.20}$$

where each term is defined by:

$$\ln \mathcal{L}^m(\{N^m(t)\}_{t \leq T}) = \int_0^T (1 - \lambda^m(s)) \, ds + \int_0^T \ln \lambda^m(s) dN^m(s). \tag{C.21}$$

This partial log-likelihood can be computed as:

$$\ln \mathcal{L}^m(\{N^m(t)\}_{t \leq T}) = T - \Lambda^m(0, T) \tag{C.22}$$

$$+ \sum_{i: T_i \leq T} \mathbf{1}_{\{Z_i = m\}} \ln \left[ \lambda_0^m(T_i) + \sum_{n=1}^{D} \sum_{j=1}^{P} \sum_{T_k^n < T_i} \alpha_{mn,j} e^{-\beta_{mn,j}(T_i - T_k^n)} \right],$$

where $\Lambda^m(0, T) = \int_0^T \lambda^m(s) ds$ is the integrated intensity. Following Ozaki (1979), we compute this in a recursive way by observing that, thanks to the exponential form of the kernel:

$$R_j^{mn}(l) = \sum_{T_k^n < T_l^m} e^{-\beta_{mn,j}(T_l^m - T_k^n)}$$

$$= \begin{cases} e^{-\beta_{mn,j}(T_l^m - T_{l-1}^m)} R_j^{mn}(l-1) + \sum_{T_{l-1}^m \leq T_k^n < T_l^m} e^{-\beta_{mn,j}(T_l^m - T_k^n)} & \text{if } m \neq n, \\ e^{-\beta_{mn,j}(T_l^m - T_{l-1}^m)} \left(1 + R_j^{mn}(l-1)\right) & \text{if } m = n. \end{cases} \tag{C.23}$$

The final expression of the partial log-likelihood may thus be written:

$$\ln \mathcal{L}^m(\{N^m(t)\}_{t \leq T}) = T - \int_0^T \lambda_0^m(s)ds - \sum_{i:T_i \leq T} \sum_{n=1}^{M} \sum_{j=1}^{P} \frac{\alpha_{mn,j}}{\beta_{mn,j}}\left(1 - e^{-\beta_{mn,j}(T-T_i)}\right)$$

$$+ \sum_{l:T_l^m \leq T} \ln\left[\lambda_0^m(T_l^m) + \sum_{n=1}^{M} \sum_{j=1}^{P} \alpha_{mn,j} R_j^{mn}(l)\right], \qquad (C.24)$$

where $R_j^{mn}(l)$ is defined with Eq. (C.23) and $R_j^{mn}(0) = 0$.

**Testing the calibration**

A general result on point processes theory states that a given non-Poisson process can be transformed into a homogeneous Poisson process by a stochastic time change. A standard monovariate version of this result and its proof can be found in (Brémaud, 1981, Chapter II, Theorem T16). Bowsher (2007) has shown that this can be generalized in a multi-dimensional setting, which provides specification tests for multi-dimensional Hawkes models. We reproduce here its result, with slightly modified notations to accommodate our notations.

**Theorem C.6** (Bowsher, 2007, Theorem 4.1) *Let $N$ be a $D$-variate point process on $\mathbb{R}_+^*$ with natural filtration $\{\mathcal{F}_t^N\}_{t \in \mathbb{R}_+}$, and $D \geq 1$. Also let $\{\mathcal{F}_t\}_{t \in \mathbb{R}_+}$ be a history of $N$ (that is, $\mathcal{F}_t^N \subseteq \mathcal{F}_t, \forall t \geq 0$), and suppose, for each $m$, that $N^m$ has the $\mathcal{F}_t$-intensity $\lambda^m$ where $\lambda^m$ satisfies $\int_0^\infty \lambda^m(s)ds = \infty$ almost surely. Define for each $m$ and all $t \geq 0$ the $\mathcal{F}_t$-stopping time $\tau^m(t)$ as the (unique) solution to*

$$\int_0^{\tau^m(t)} \lambda^m(u)du = t. \qquad (C.25)$$

*Then the $D$ point processes $\{\tilde{N}^m\}_{m=1,\ldots,D}$ defined by $\tilde{N}^m(t) = N^m(\tau^m(t))$, $\forall t \geq 0$, are independent Poisson processes with unit intensity. Furthermore, the durations of each Poisson process $\tilde{N}^m$ are given by*

$$\Lambda^m(T_{i-1}^m, T_i^m) = \int_{T_{i-1}^m}^{T_i^m} \lambda^m(s)ds, \quad \forall i \geq 2. \qquad (C.26)$$

Let us compute the integrated intensity of the $m$-th coordinate of a multidimensional Hawkes process between two consecutive events $T_{i-1}^m$ and $T_i^m$ of type $m$:

$$\Lambda^m(T_{i-1}^m, T_i^m) = \int_{T_{i-1}^m}^{T_i^m} \lambda^m(s)ds$$

$$= \int_{T_{i-1}^m}^{T_i^m} \lambda_0^m(s)ds + \sum_{n=1}^{D}\sum_{j=1}^{P}\sum_{T_k^n < T_{i-1}^m} \frac{\alpha_{mn,j}}{\beta_{mn,j}}\left[e^{-\beta_{mn,j}(T_{i-1}^m - T_k^n)} - e^{-\beta_{mn,j}(T_i^m - T_k^n)}\right]$$

$$+ \sum_{n=1}^{D}\sum_{j=1}^{P}\sum_{T_{i-1}^m \leq T_k^n < T_i^m} \frac{\alpha_{mn,j}}{\beta_{mn,j}}\left[1 - e^{-\beta_{mn,j}(T_i^m - T_k^n)}\right]. \tag{C.27}$$

As in the log-likelihood computation, following Ozaki (1979), we observe that:

$$A_j^{mn}(i-1) = \sum_{T_k^n < T_{i-1}^m} e^{-\beta_{mn,j}(T_{i-1}^m - T_k^n)} \tag{C.28}$$

$$= e^{-\beta_{mn,j}(T_{i-1}^m - T_{i-2}^m)} A_j^{mn}(i-2) + \sum_{T_{i-2}^m \leq T_k^n < T_{i-1}^m} e^{-\beta_{mn,j}(T_{i-1}^m - T_k^n)},$$

so that the integrated density can be written for all $i \geq 2$:

$$\Lambda^m(T_{i-1}^m, T_i^m) = \int_{T_{i-1}^m}^{T_i^m} \lambda_0^m(s)ds + \sum_{n=1}^{D}\sum_{j=1}^{P} \frac{\alpha_{mn,j}}{\beta_{mn,j}}\Bigg[A_j^{mn}(i-1)\left(1 - e^{-\beta_{mn,j}(T_i^m - T_{i-1}^m)}\right)$$

$$+ \sum_{T_{i-1}^m \leq T_k^n < T_i^m} \left(1 - e^{-\beta_j^{mn}(T_i^m - T_k^n)}\right)\Bigg], \tag{C.29}$$

where $A_j^{mn}$ is defined as in Eq. (C.28) with for all $j = 1, \ldots, P, A_j^{mn}(0) = 0$.

Hence, simply following the method in Bowsher (2007), we can easily define tests to check the goodness-of-fit of a Hawkes model to some empirical data. Since, the integrated intensity $\Lambda^m(T_{i-1}^m, T_i^m)$ is a time interval of a homogeneous Poisson Process, we can test for each $m = 1, \ldots, D$: (i) whether the variables $\left(\Lambda^m(T_{i-1}^m, T_i^m)\right)_{i \geq 2}$ are exponentially distributed; (ii) whether the variables $\left((\Lambda^m(T_{i-1}^m, T_i^m)\right)_{i \geq 2}$ are independent.

## C.2 Ergodic Theory for Markov Processes

The ergodicity of a Markov process is a fundamental notion related to the possible identification of averages over time or space. Loosely speaking, ergodicity characterizes those processes whose sample paths visit the state space in a uniform (with respect to some measure) manner. Conditions for ergodicity and convergence towards an invariant measure are provided by the theory of *stochastic stability*, for which we refer to Meyn and Tweedie (2009) and simply recall some important and useful results.

### C.2.1 Stochastic stability

Let $(Q^t)_{t \geq 0}$ be the transition probability function of a Markov process at time $t$, that is

$$Q^t(\mathbf{x}, E) := \mathbb{P}\left[\mathbf{X}(t) \in E | \mathbf{X}(0) = \mathbf{x}\right], \; t \in \mathbb{R}_+, \mathbf{x} \in \mathcal{S}, E \subset \mathcal{S}, \tag{C.30}$$

where $\mathcal{S}$ is the state space of the process. An aperiodic, irreducible Markov process is *ergodic* if an invariant probability measure $\pi$ exists and

$$\lim_{t \to \infty} \|Q^t(\mathbf{x}, .) - \pi(.)\| = 0, \forall \mathbf{x} \in \mathcal{S}, \tag{C.31}$$

where $\|.\|$ designates for a signed measure $\nu$ the *total variation norm* defined as

$$\|\nu\| := \sup_{f : |f| < 1} |\nu(f)| = \sup_{E \in \mathcal{B}(\mathcal{S})} \nu(E) - \inf_{E \in \mathcal{B}(\mathcal{S})} \nu(E). \tag{C.32}$$

In (C.32), $\mathcal{B}(\mathcal{S})$ is the Borel $\sigma$-field generated by $\mathcal{S}$, and for a measurable function $f$ on $\mathcal{S}$, $\nu(f) := \int_{\mathcal{S}} f d\nu$.

*V-uniform ergodicity.* A Markov process is said *V-uniformly ergodic* if there exists a coercive function $V > 1$, an invariant distribution $\pi$, and constants $r$, $0 < r < 1$, and $R < \infty$ such that

$$\|Q^t(\mathbf{x}, .) - \pi(.)\| \leq R r^t V(\mathbf{x}), \mathbf{x} \in \mathcal{S}, t \in \mathbb{R}_+. \tag{C.33}$$

V−uniform ergodicity is studied *via* the infinitesimal generator of the Markov process. Indeed, it is shown in Meyn and Tweedie (2009, 1993) that it is equivalent to the existence of a coercive function $V$ satisfying the Lyapunov-type condition

$$\mathcal{L}V(\mathbf{x}) \leq -\beta V(\mathbf{x}) + \gamma \mathbf{1}_C, \; \text{(Geometric drift condition)} \tag{C.34}$$

for some positive constants $\beta$ and $\gamma$, and where **C** is a **petite** set. (Theorems 6.1 and 7.1 in Meyn and Tweedie (1993).) Condition (C.34) says that the larger $V(\mathbf{X}(t))$, the stronger **X** is pulled back towards the center of the state space $\mathcal{S}$. We refer to Meyn and Tweedie (2009) for further details.

Due to the topology of the state space, a Lyapunov function is often obtained under the form

$$\mathcal{L}V(\mathbf{x}) \leq -\beta V(\mathbf{x}) + \gamma \mathbf{1}_{\mathbf{K}}, \quad \text{(Geometric drift condition)}, \tag{C.35}$$

where **K** is a compact set rather than a petite set. Hence, it is important to have criteria showing that compact sets are indeed petite sets. Such criteria are provided in Chapter 6 of Meyn and Tweedie (2009), and can be obtained directly in specific examples. For instance, the case of countable state space is well-known and covers the zero-intelligence model in Chapter 6; as for Hawkes processes, such a result is proven in Zheng et al. (2014), and the proof given there easily extends to the case of a Hawkes process-driven limit order book.

### C.2.2 The Ergodic Theorem and Martingale Convergence Theorem

The Ergodic Theorem for Markov processes states the following:

> **Theorem C.7** (Meyn and Tweedie (2009) Maruyama and Tanaka (1959) Cattiaux et al. (2012)) *Let X be an ergodic Markov process. Denote by $\Pi$ its unique invariant probability measure, and let H be in $L^1(\Pi(dX))$. Then, there holds:*
>
> $$\lim_{t \to +\infty} \frac{1}{t} \int_0^t G(X(t)) dt \stackrel{a.s.}{=} \int G(x) \Pi(dx).$$

The Martingale Convergence Theorem states a general invariance principle for conveniently rescaled martingales under minimal assumptions of convergence for the quadratic variation and jump sizes. We quote below the version that is used in this book.

> **Theorem C.8** (Theorem 7:1:4 in Ethier and Kurtz (2005), Theorem 2.1 in Whitt (2007)) *For $n \geq 1$, let $\mathbf{M}_n \equiv (M_{n,1}, ..., M_{n,k})$ be a local martingale in the Skorohod space $D^k$ with respect to a filtration $(\mathcal{F}_{n,t} : t \geq 0)$, satisfying $\mathbf{M}_n(0) = 0$. Let $\mathbf{C} \equiv (C_{ij})$ be a covariance matrix, i.e. a nonnegative-definite symmetric matrix of real numbers.*
>
> **Assume the following:** *$\mathbf{M}_n$ is locally square-integrable. The expected value of the maximum jump in the predictable quadratic variation $\langle M_{n,i}, M_{n,j} \rangle$ and of the maximum squared jump of $\mathbf{M}_n$ are asymptotically negligible. Furthermore*
>
> $$\langle M_{n,i}, M_{n,j} \rangle (t) \Rightarrow c_{ij}(t) \text{ in } \mathbb{R} \text{ as } n \to \infty \tag{C.36}$$
>
> *for each $t > 0$ and for each pair $i, j$.*

***Conclusion:***

$$\mathbf{M}_n \Rightarrow \mathbf{M} \text{ in } D^k \text{ as } n \to \infty, \tag{C.37}$$

where $\mathbf{M}$ *is a k-dimensional Wiener process with mean vector* $E(\mathbf{M}(t)) = \mathbf{0}$ *and covariance matrix* $E(\mathbf{M}(t)\mathbf{M}(t)^{tr}) = \mathbf{C}t$.

# APPENDIX D

# Comparison of Various Prediction Methods

This appendix presents the numerical results for the various prediction methods presented and back-tested in Chapter 10.

## D.1 Results for the Binary Classification

**Table D.1** The quality of the binary prediction: 1-minute prediction AUC and accuracy per stock

|  | Order book imbalance | | Flow quantity | | Past return | |
|---|---|---|---|---|---|---|
| Stock | AUC | Accuracy | AUC | Accuracy | AUC | Accuracy |
| INTERBREW | 0.54 | 0.54 | 0.53 | 0.53 | 0.51 | 0.51 |
| AIR LIQUIDE | 0.56 | 0.56 | 0.53 | 0.53 | 0.51 | 0.51 |
| ALLIANZ | 0.61 | 0.61 | 0.51 | 0.51 | 0.53 | 0.53 |
| ASML Holding NV | 0.56 | 0.56 | 0.53 | 0.53 | 0.51 | 0.51 |
| BASF AG | 0.54 | 0.54 | 0.52 | 0.52 | 0.50 | 0.50 |
| BAYER AG | 0.54 | 0.54 | 0.53 | 0.53 | 0.51 | 0.51 |
| BBVARGENTARIA | 0.54 | 0.54 | 0.53 | 0.53 | 0.51 | 0.51 |
| BAY MOT WERKE | 0.54 | 0.54 | 0.53 | 0.53 | 0.51 | 0.51 |
| DANONE | 0.56 | 0.56 | 0.53 | 0.53 | 0.51 | 0.51 |
| BNP PARIBAS | 0.53 | 0.53 | 0.52 | 0.52 | 0.51 | 0.51 |
| CARREFOUR | 0.55 | 0.55 | 0.53 | 0.53 | 0.51 | 0.51 |
| CRH PLC IRLANDE | 0.62 | 0.62 | 0.58 | 0.58 | 0.53 | 0.53 |
| AXA | 0.55 | 0.55 | 0.51 | 0.51 | 0.52 | 0.52 |
| DAIMLER CHRYSLER | 0.54 | 0.54 | 0.53 | 0.53 | 0.51 | 0.51 |

*Contd...*

|  | Order book imbalance | | Flow quantity | | Past return | |
|---|---|---|---|---|---|---|
| Stock | AUC | Accuracy | AUC | Accuracy | AUC | Accuracy |
| DEUTSCHE BANK AG | 0.53 | 0.53 | 0.52 | 0.52 | 0.51 | 0.51 |
| VINCI | 0.54 | 0.54 | 0.53 | 0.53 | 0.51 | 0.51 |
| DEUTSCHE TELEKOM | 0.56 | 0.56 | 0.52 | 0.52 | 0.51 | 0.51 |
| ESSILOR INTERNATIONAL | 0.56 | 0.56 | 0.54 | 0.54 | 0.50 | 0.50 |
| ENEL | 0.63 | 0.63 | 0.51 | 0.51 | 0.55 | 0.55 |
| ENI | 0.64 | 0.64 | 0.51 | 0.51 | 0.56 | 0.56 |
| E.ON AG | 0.58 | 0.58 | 0.51 | 0.51 | 0.51 | 0.51 |
| TOTAL | 0.54 | 0.54 | 0.52 | 0.52 | 0.51 | 0.51 |
| GENERALI ASSIC | 0.62 | 0.62 | 0.50 | 0.50 | 0.54 | 0.54 |
| SOCIETE GENERALE | 0.52 | 0.52 | 0.51 | 0.51 | 0.51 | 0.51 |
| GDF SUEZ | 0.56 | 0.56 | 0.52 | 0.52 | 0.50 | 0.50 |
| IBERDROLA I | 0.56 | 0.56 | 0.54 | 0.54 | 0.51 | 0.51 |
| ING | 0.53 | 0.53 | 0.53 | 0.53 | 0.51 | 0.51 |
| INTESABCI | 0.60 | 0.60 | 0.51 | 0.51 | 0.53 | 0.53 |
| INDITEX | 0.59 | 0.59 | 0.55 | 0.55 | 0.50 | 0.50 |
| LVMH | 0.59 | 0.59 | 0.52 | 0.52 | 0.52 | 0.52 |
| MUNICH RE | 0.58 | 0.58 | 0.52 | 0.52 | 0.51 | 0.51 |
| LOREAL | 0.60 | 0.60 | 0.53 | 0.53 | 0.52 | 0.52 |
| PHILIPS ELECTR. | 0.56 | 0.56 | 0.55 | 0.55 | 0.50 | 0.50 |
| REPSOL | 0.57 | 0.57 | 0.54 | 0.54 | 0.51 | 0.51 |
| RWE ST | 0.54 | 0.54 | 0.53 | 0.53 | 0.51 | 0.51 |
| BANCO SAN CENTRAL HISPANO | 0.54 | 0.54 | 0.53 | 0.53 | 0.51 | 0.51 |
| SANOFI | 0.54 | 0.54 | 0.53 | 0.53 | 0.50 | 0.50 |
| SAP AG | 0.54 | 0.54 | 0.52 | 0.52 | 0.51 | 0.51 |
| SAINT GOBAIN | 0.54 | 0.54 | 0.53 | 0.53 | 0.51 | 0.51 |
| SIEMENS AG | 0.54 | 0.54 | 0.53 | 0.53 | 0.51 | 0.51 |
| SCHNEIDER ELECTRIC SA | 0.54 | 0.54 | 0.52 | 0.52 | 0.51 | 0.51 |
| TELEFONICA | 0.59 | 0.59 | 0.53 | 0.53 | 0.51 | 0.51 |
| UNICREDIT SPA | 0.57 | 0.57 | 0.50 | 0.50 | 0.52 | 0.52 |
| UNILEVER CERT | 0.56 | 0.56 | 0.52 | 0.52 | 0.51 | 0.51 |
| VIVENDI UNIVERSAL | 0.57 | 0.57 | 0.53 | 0.53 | 0.51 | 0.51 |
| VOLKSWAGEN | 0.57 | 0.57 | 0.52 | 0.52 | 0.51 | 0.51 |

**Table D.2** The quality of the binary prediction: The daily gain average and standard deviation for the 1-minute prediction (without trading costs)

| Stock | Order book imbalance | | Flow quantity | | Past return | |
|---|---|---|---|---|---|---|
| | $\overline{Gain}$ | $\sigma(Gain)$ | $\overline{Gain}$ | $\sigma(Gain)$ | $\overline{Gain}$ | $\sigma(Gain)$ |
| INTERBREW | 1388 | 1201 | 1107 | 1308 | 174 | 1264 |
| AIR LIQUIDE | 1603 | 1112 | 996 | 1005 | 169 | 936 |
| ALLIANZ | 2775 | 1219 | 221 | 1107 | 638 | 1175 |
| ASML Holding NV | 1969 | 1278 | 1244 | 1316 | 190 | 1419 |
| BASF AG | 1156 | 1102 | 921 | 1311 | 2 | 1185 |
| BAYER AG | 1269 | 1055 | 1142 | 1251 | 289 | 1296 |
| BBVARGENTARIA | 1954 | 1537 | 1866 | 1700 | 595 | 1934 |
| BAY MOT WERKE | 1330 | 1219 | 1240 | 1325 | 347 | 1394 |
| DANONE | 1591 | 993 | 958 | 1143 | 231 | 1196 |
| BNP PARIBAS | 1120 | 1608 | 831 | 1620 | 526 | 1911 |
| CARREFOUR | 1878 | 1572 | 1461 | 1601 | 600 | 1665 |
| CRH PLC IRLANDE | 4144 | 1881 | 2853 | 1691 | 1496 | 1542 |
| AXA | 2003 | 1373 | 674 | 1428 | 582 | 1603 |
| DAIMLER CHRYSLER | 1380 | 1275 | 1130 | 1228 | 208 | 1390 |
| DEUTSCHE BANK AG | 1251 | 1372 | 905 | 1405 | 310 | 1672 |
| VINCI | 1410 | 1113 | 1252 | 1211 | 376 | 1113 |
| DEUTSCHE TELEKOM | 1586 | 1416 | 848 | 1196 | 308 | 1298 |
| ESSILOR INTERNATIONAL | 1762 | 1315 | 1523 | 1295 | 12 | 1281 |
| ENEL | 3723 | 1655 | 295 | 1384 | 1219 | 1307 |
| ENI | 2996 | 1185 | 321 | 1161 | 1109 | 1201 |
| E.ON AG | 2245 | 1193 | 481 | 1722 | 323 | 1445 |
| TOTAL | 1256 | 956 | 831 | 977 | 326 | 950 |
| GENERALI ASSIC | 3977 | 1764 | 177 | 1324 | 1210 | 1577 |
| SOCIETE GENERALE | 1195 | 1763 | 853 | 1896 | 643 | 2060 |
| GDF SUEZ | 2031 | 1227 | 934 | 1389 | 156 | 1355 |
| IBERDROLA I | 2220 | 1433 | 1626 | 1514 | 566 | 1403 |
| ING | 1511 | 1564 | 1493 | 1491 | 217 | 1720 |
| INTESABCI | 4019 | 1911 | 153 | 1787 | 1048 | 1954 |
| INDITEX | 2481 | 1452 | 1742 | 1525 | 145 | 1344 |

*Contd...*

|  | Order book imbalance | | Flow quantity | | Past return | |
|---|---|---|---|---|---|---|
| Stock | $\overline{Gain}$ | $\sigma(Gain)$ | $\overline{Gain}$ | $\sigma(Gain)$ | $\overline{Gain}$ | $\sigma(Gain)$ |
| LVMH | 2445 | 1220 | 533 | 1148 | 613 | 1267 |
| MUNICH RE | 1895 | 1107 | 791 | 1485 | 194 | 1006 |
| LOREAL | 2367 | 1109 | 894 | 1242 | 438 | 1220 |
| PHILIPS ELECTR. | 1978 | 1173 | 1670 | 1565 | 182 | 1251 |
| REPSOL | 2694 | 1451 | 1700 | 1607 | 292 | 1558 |
| RWE ST | 1323 | 1348 | 1475 | 1880 | 307 | 1747 |
| BANCO SAN CENTRAL HISPANO | 1717 | 1535 | 1393 | 1577 | 383 | 1684 |
| SANOFI | 1368 | 1040 | 1118 | 1123 | 107 | 1190 |
| SAP AG | 1225 | 1022 | 939 | 1071 | 117 | 1084 |
| SAINT GOBAIN | 1612 | 1359 | 1209 | 1449 | 455 | 1607 |
| SIEMENS AG | 1108 | 983 | 967 | 1196 | 164 | 1124 |
| SCHNEIDER ELECTRIC SA | 1419 | 1294 | 1014 | 1275 | 379 | 1436 |
| TELEFONICA | 2694 | 1267 | 1156 | 1341 | 290 | 1194 |
| UNICREDIT SPA | 3039 | 2025 | 382 | 1850 | 683 | 2002 |
| UNILEVER CERT | 1402 | 766 | 551 | 860 | 222 | 949 |
| VIVENDI UNIVERSAL | 2142 | 1223 | 1114 | 1391 | 244 | 1326 |
| VOLKSWAGEN | 2044 | 1440 | 1165 | 1397 | 225 | 1359 |

**Table D.3** The quality of the binary prediction: The daily gain average and standard deviation for the 1-minute prediction (with trading costs)

|  | Order book imbalance | | Flow quantity | | Past return | |
|---|---|---|---|---|---|---|
| Stock | $\overline{Gain}$ | $\sigma(Gain)$ | $\overline{Gain}$ | $\sigma(Gain)$ | $\overline{Gain}$ | $\sigma(Gain)$ |
| INTERBREW | −191 | 1189 | −788 | 1325 | −1222 | 1531 |
| AIR LIQUIDE | 81 | 1112 | −980 | 1057 | −1211 | 1164 |
| ALLIANZ | 1141 | 1063 | −1199 | 1309 | −952 | 1162 |
| ASML Holding NV | 370 | 1179 | −697 | 1335 | −1301 | 1574 |
| BASF AG | −422 | 1064 | −955 | 1338 | −1298 | 1558 |
| BAYER AG | −363 | 1002 | −734 | 1249 | −1122 | 1503 |
| BBVARGENTARIA | 303 | 1477 | −58 | 1681 | −910 | 2027 |
| BAY MOT WERKE | −260 | 1176 | −530 | 1263 | −1256 | 1510 |

*Contd...*

|  | Order book imbalance | | Flow quantity | | Past return | |
|---|---|---|---|---|---|---|
| Stock | $\overline{Gain}$ | $\sigma(Gain)$ | $\overline{Gain}$ | $\sigma(Gain)$ | $\overline{Gain}$ | $\sigma(Gain)$ |
| DANONE | −40 | 963 | −906 | 1164 | −1246 | 1369 |
| BNP PARIBAS | −402 | 1596 | −1022 | 1618 | −1115 | 1998 |
| CARREFOUR | 251 | 1486 | −492 | 1606 | −975 | 1690 |
| CRH PLC IRLANDE | 2971 | 1714 | 934 | 1612 | −27 | 1549 |
| AXA | 313 | 1299 | −1064 | 1488 | −1152 | 1560 |
| DAIMLER CHRYSLER | −231 | 1243 | −748 | 1235 | −1206 | 1529 |
| DEUTSCHE BANK AG | −394 | 1368 | −959 | 1423 | −1277 | 1819 |
| VINCI | −170 | 1072 | −656 | 1224 | −1093 | 1324 |
| DEUTSCHE TELEKOM | 50 | 1407 | −949 | 1225 | −1128 | 1516 |
| ESSILOR INTERNATIONAL | 185 | 1265 | −389 | 1296 | −1104 | 1575 |
| ENEL | 2151 | 1456 | −1069 | 1610 | −329 | 1198 |
| ENI | 1513 | 971 | −1136 | 1375 | −281 | 1046 |
| E.ON AG | 583 | 1096 | −1108 | 1887 | −1047 | 1592 |
| TOTAL | −362 | 934 | −1058 | 1024 | −1278 | 1206 |
| GENERALI ASSIC | 2369 | 1565 | −1403 | 1539 | −484 | 1490 |
| SOCIETE GENERALE | −405 | 1718 | −846 | 1901 | −968 | 2002 |
| GDF SUEZ | 402 | 1140 | −951 | 1438 | −1249 | 1513 |
| IBERDROLA I | 762 | 1332 | −312 | 1503 | −1094 | 1475 |
| ING | −186 | 1519 | −450 | 1470 | −1186 | 1890 |
| INTESABCI | 2333 | 1715 | −1081 | 1822 | −517 | 1820 |
| INDITEX | 1110 | 1375 | −195 | 1535 | −1155 | 1457 |
| LVMH | 831 | 1119 | −1183 | 1296 | −928 | 1235 |
| MUNICH RE | 366 | 1011 | −1019 | 1490 | −1260 | 1177 |
| LOREAL | 816 | 985 | −797 | 1274 | −982 | 1236 |
| PHILIPS ELECTR. | 377 | 1113 | −272 | 1575 | −1255 | 1490 |
| REPSOL | 1233 | 1308 | −184 | 1585 | −1188 | 1713 |
| RWE ST | −182 | 1251 | −399 | 1864 | −1122 | 1960 |
| BANCO SAN CENTRAL HISPANO | 205 | 1431 | −492 | 1566 | −1064 | 1822 |
| SANOFI | −279 | 998 | −720 | 1127 | −1382 | 1454 |
| SAP AG | −340 | 1000 | −944 | 1093 | −1428 | 1277 |

*Contd...*

|  | Order book imbalance | | Flow quantity | | Past return | |
|---|---|---|---|---|---|---|
| Stock | $\overline{Gain}$ | $\sigma(Gain)$ | $\overline{Gain}$ | $\sigma(Gain)$ | $\overline{Gain}$ | $\sigma(Gain)$ |
| SAINT GOBAIN | −48 | 1326 | −694 | 1463 | −1060 | 1655 |
| SIEMENS AG | −472 | 966 | −898 | 1209 | −1353 | 1363 |
| SCHNEIDER ELECTRIC SA | −162 | 1263 | −872 | 1296 | −1339 | 1493 |
| TELEFONICA | 1124 | 1130 | −686 | 1342 | −1044 | 1257 |
| UNICREDIT SPA | 1434 | 1940 | −896 | 1953 | −738 | 2067 |
| UNILEVER CERT | −253 | 730 | −1246 | 938 | −1344 | 1142 |
| VIVENDI UNIVERSAL | 547 | 1113 | −804 | 1386 | −1186 | 1452 |
| VOLKSWAGEN | 446 | 1373 | −785 | 1408 | −979 | 1584 |

## D.2 Results for the Four-class Classification

**Table D.4** The quality of the 4-class prediction: 1-minute prediction AUC and accuracy per stock

|  | Order book imbalance | | Flow quantity | | Past return | |
|---|---|---|---|---|---|---|
| Stock | AUC | Accuracy | AUC | Accuracy | AUC | Accuracy |
| INTERBREW | 0.58 | 0.59 | 0.50 | 0.42 | 0.50 | 0.50 |
| AIR LIQUIDE | 0.71 | 0.72 | nan | nan | 0.50 | 0.58 |
| ALLIANZ | 0.69 | 0.69 | 0.50 | 0.54 | 0.61 | 0.61 |
| ASML Holding NV | 0.60 | 0.60 | 0.50 | 0.54 | 0.48 | 0.48 |
| BASF AG | 0.60 | 0.60 | nan | nan | 0.49 | 0.50 |
| BAYER AG | 0.53 | 0.55 | 0.50 | 0.59 | 0.50 | 0.56 |
| BBVARGENTARIA | 0.57 | 0.57 | 0.55 | 0.55 | 0.55 | 0.56 |
| BAY MOT WERKE | 0.57 | 0.58 | 0.55 | 0.55 | 0.54 | 0.55 |
| DANONE | 0.60 | 0.60 | nan | nan | 0.58 | 0.58 |
| BNP PARIBAS | 0.58 | 0.59 | 0.50 | 0.50 | 0.52 | 0.53 |
| CARREFOUR | 0.59 | 0.60 | 0.50 | 0.56 | 0.56 | 0.56 |
| CRH PLC IRLANDE | 0.70 | 0.70 | 0.64 | 0.64 | 0.55 | 0.56 |
| AXA | 0.58 | 0.60 | nan | nan | 0.56 | 0.56 |
| DAIMLER CHRYSLER | 0.57 | 0.57 | 0.50 | 0.51 | 0.54 | 0.54 |
| DEUTSCHE BANK AG | 0.55 | 0.55 | 0.54 | 0.56 | 0.52 | 0.52 |
| VINCI | 0.60 | 0.60 | 0.55 | 0.56 | 0.56 | 0.56 |

*Contd...*

|  | Order book imbalance | | Flow quantity | | Past return | |
|---|---|---|---|---|---|---|
| Stock | AUC | Accuracy | AUC | Accuracy | AUC | Accuracy |
| DEUTSCHE TELEKOM | 0.71 | 0.72 | nan | nan | 0.51 | 0.51 |
| ESSILOR INTERNATIONAL | 0.60 | 0.60 | 0.50 | 0.55 | 0.52 | 0.56 |
| ENEL | 0.73 | 0.73 | nan | nan | 0.57 | 0.60 |
| ENI | 0.76 | 0.76 | nan | nan | 0.61 | 0.61 |
| E.ON AG | 0.64 | 0.64 | nan | nan | 0.53 | 0.53 |
| TOTAL | 0.54 | 0.59 | nan | nan | 0.50 | 0.46 |
| GENERALI ASSIC | 0.68 | 0.68 | nan | nan | 0.60 | 0.60 |
| SOCIETE GENERALE | 0.55 | 0.56 | 0.50 | 0.54 | 0.52 | 0.54 |
| GDF SUEZ | 0.62 | 0.62 | nan | nan | 0.53 | 0.53 |
| IBERDROLA I | 0.63 | 0.63 | 0.56 | 0.56 | 0.57 | 0.57 |
| ING | 0.55 | 0.55 | 0.54 | 0.55 | 0.52 | 0.55 |
| INTESABCI | 0.67 | 0.67 | nan | nan | 0.58 | 0.58 |
| INDITEX | 0.68 | 0.68 | 0.58 | 0.58 | 0.55 | 0.55 |
| LVMH | 0.65 | 0.66 | nan | nan | 0.58 | 0.58 |
| MUNICH RE | 0.66 | 0.66 | 0.55 | 0.55 | 0.54 | 0.54 |
| LOREAL | 0.67 | 0.67 | nan | nan | 0.58 | 0.58 |
| PHILIPS ELECTR. | 0.61 | 0.62 | 0.50 | 0.51 | 0.52 | 0.54 |
| REPSOL | 0.63 | 0.63 | 0.53 | 0.58 | 0.57 | 0.57 |
| RWE ST | 0.58 | 0.58 | 0.53 | 0.55 | 0.52 | 0.52 |
| BANCO SAN CENTRAL HISPANO | 0.57 | 0.56 | 0.52 | 0.51 | 0.58 | 0.58 |
| SANOFI | 0.60 | 0.60 | nan | nan | 0.50 | 0.60 |
| SAP AG | 0.52 | 0.61 | 0.50 | 0.56 | 0.52 | 0.54 |
| SAINT GOBAIN | 0.56 | 0.58 | 0.54 | 0.58 | 0.54 | 0.55 |
| SIEMENS AG | 0.56 | 0.61 | 0.55 | 0.56 | 0.59 | 0.59 |
| SCHNEIDER ELECTRIC SA | 0.57 | 0.58 | nan | nan | 0.56 | 0.57 |
| TELEFONICA | 0.68 | 0.68 | 0.53 | 0.57 | 0.56 | 0.56 |
| UNICREDIT SPA | 0.64 | 0.65 | 0.50 | 0.54 | 0.57 | 0.57 |
| UNILEVER CERT | 0.50 | 0.63 | nan | nan | nan | nan |
| VIVENDI UNIVERSAL | 0.63 | 0.63 | nan | nan | 0.51 | 0.52 |
| VOLKSWAGEN | 0.62 | 0.62 | 0.49 | 0.49 | 0.52 | 0.53 |

**Table D.5** The quality of the 4-class prediction: The daily gain average and standard deviation for the 1-minute prediction (without trading costs)

| Stock | Order book imbalance | | Flow quantity | | Past return | |
|---|---|---|---|---|---|---|
| | $\overline{Gain}$ | $\sigma(Gain)$ | $\overline{Gain}$ | $\sigma(Gain)$ | $\overline{Gain}$ | $\sigma(Gain)$ |
| INTERBREW | 137 | 388 | −6 | 98 | 4 | 131 |
| AIR LIQUIDE | 306 | 577 | 0 | 0 | 3 | 42 |
| ALLIANZ | 1363 | 779 | 4 | 47 | 68 | 276 |
| ASML Holding NV | 440 | 651 | 5 | 63 | −2 | 132 |
| BASF AG | 87 | 287 | 0 | 0 | −2 | 48 |
| BAYER AG | 21 | 128 | 14 | 137 | 14 | 99 |
| BBVARGENTARIA | 390 | 665 | 273 | 669 | 208 | 582 |
| BAY MOT WERKE | 107 | 281 | 47 | 276 | 52 | 238 |
| DANONE | 168 | 366 | 0 | 0 | 4 | 47 |
| BNP PARIBAS | 171 | 428 | 3 | 66 | 44 | 453 |
| CARREFOUR | 486 | 715 | 11 | 139 | 136 | 469 |
| CRH PLC IRLANDE | 2534 | 1240 | 1364 | 1077 | 202 | 560 |
| AXA | 594 | 786 | 0 | 0 | 55 | 320 |
| DAIMLER CHRYSLER | 93 | 289 | 2 | 24 | 16 | 191 |
| DEUTSCHE BANK AG | 34 | 224 | 38 | 212 | 12 | 291 |
| VINCI | 154 | 451 | 13 | 111 | 27 | 147 |
| DEUTSCHE TELEKOM | 488 | 827 | 0 | 0 | 3 | 66 |
| ESSILOR INTERNATIONAL | 351 | 596 | 17 | 164 | 10 | 106 |
| ENEL | 2219 | 1056 | 0 | 0 | 193 | 503 |
| ENI | 2000 | 773 | 0 | 0 | 110 | 300 |
| E.ON AG | 651 | 680 | 0 | 0 | 10 | 168 |
| TOTAL | 10 | 93 | 0 | 0 | 1 | 38 |
| GENERALI ASSIC | 2520 | 1420 | 0 | 0 | 249 | 756 |
| SOCIETE GENERALE | 184 | 503 | 2 | 25 | 56 | 410 |
| GDF SUEZ | 504 | 692 | 0 | 0 | 21 | 171 |
| IBERDROLA I | 738 | 951 | 155 | 512 | 115 | 409 |
| ING | 109 | 373 | 59 | 296 | 7 | 138 |
| INTESABCI | 2512 | 1248 | 0 | 0 | 185 | 731 |

*Contd...*

| Stock | Order book imbalance | | Flow quantity | | Past return | |
|---|---|---|---|---|---|---|
| | $\overline{Gain}$ | $\sigma(Gain)$ | $\overline{Gain}$ | $\sigma(Gain)$ | $\overline{Gain}$ | $\sigma(Gain)$ |
| INDITEX | 1039 | 914 | 151 | 587 | 44 | 223 |
| LVMH | 930 | 847 | 0 | 0 | 64 | 277 |
| MUNICH RE | 370 | 533 | 26 | 145 | 3 | 50 |
| LOREAL | 800 | 674 | 0 | 0 | 22 | 112 |
| PHILIPS ELECTR. | 440 | 613 | 6 | 94 | 11 | 116 |
| REPSOL | 1234 | 1013 | 142 | 445 | 110 | 555 |
| RWE ST | 192 | 556 | 85 | 380 | 29 | 364 |
| BANCO SAN CENTRAL HISPANO | 228 | 501 | 4 | 158 | 168 | 635 |
| SANOFI | 26 | 127 | 0 | 0 | 6 | 90 |
| SAP AG | 50 | 196 | 24 | 187 | 6 | 200 |
| SAINT GOBAIN | 210 | 519 | 30 | 186 | 88 | 362 |
| SIEMENS AG | 26 | 139 | 31 | 198 | 28 | 162 |
| SCHNEIDER ELECTRIC SA | 123 | 434 | 0 | 0 | 37 | 214 |
| TELEFONICA | 1402 | 825 | 36 | 232 | 34 | 205 |
| UNICREDIT SPA | 1316 | 1393 | 17 | 197 | 247 | 835 |
| UNILEVER CERT | 16 | 104 | 0 | 0 | 0 | 0 |
| VIVENDI UNIVERSAL | 583 | 826 | 0 | 0 | 5 | 141 |
| VOLKSWAGEN | 530 | 745 | −0 | 78 | 1 | 215 |

**Table D.6** The quality of the 4-class prediction: The daily gain average and standard deviation for the 1-minute prediction (with trading costs)

| Stock | Order book imbalance | | Flow quantity | | Past return | |
|---|---|---|---|---|---|---|
| | $\overline{Gain}$ | $\sigma(Gain)$ | $\overline{Gain}$ | $\sigma(Gain)$ | $\overline{Gain}$ | $\sigma(Gain)$ |
| INTERBREW | 22 | 263 | −9 | 150 | −38 | 183 |
| AIR LIQUIDE | 128 | 329 | 0 | 0 | 1 | 31 |
| ALLIANZ | 586 | 559 | −1 | 16 | 8 | 194 |
| ASML Holding NV | 125 | 408 | −0 | 32 | −25 | 168 |
| BASF AG | 15 | 189 | 0 | 0 | −7 | 51 |
| BAYER AG | −14 | 105 | 1 | 86 | −2 | 77 |
| BBVARGENTARIA | 107 | 507 | 31 | 465 | 16 | 474 |

*Contd...*

|  | Order book imbalance | | Flow quantity | | Past return | |
|---|---|---|---|---|---|---|
| Stock | $\overline{Gain}$ | $\sigma(Gain)$ | $\overline{Gain}$ | $\sigma(Gain)$ | $\overline{Gain}$ | $\sigma(Gain)$ |
| BAY MOT WERKE | −1 | 193 | 1 | 199 | −12 | 184 |
| DANONE | 21 | 210 | 0 | 0 | −4 | 42 |
| BNP PARIBAS | 34 | 271 | −12 | 126 | −65 | 481 |
| CARREFOUR | 116 | 506 | −8 | 131 | 18 | 362 |
| CRH PLC IRLANDE | 1848 | 1102 | 518 | 844 | 18 | 442 |
| AXA | 174 | 550 | 0 | 0 | −23 | 274 |
| DAIMLER CHRYSLER | −7 | 245 | −1 | 14 | −32 | 199 |
| DEUTSCHE BANK AG | −23 | 204 | 8 | 122 | −33 | 311 |
| VINCI | 38 | 281 | −3 | 73 | −5 | 111 |
| DEUTSCHE TELEKOM | 241 | 526 | 0 | 0 | −10 | 79 |
| ESSILOR INTERNATIONAL | 88 | 388 | −14 | 157 | −4 | 91 |
| ENEL | 1338 | 881 | 0 | 0 | −18 | 443 |
| ENI | 1082 | 613 | 0 | 0 | −25 | 211 |
| E.ON AG | 185 | 475 | 0 | 0 | −22 | 173 |
| TOTAL | −5 | 72 | 0 | 0 | −3 | 49 |
| GENERALI ASSIC | 1518 | 1179 | 0 | 0 | 58 | 636 |
| SOCIETE GENERALE | 2 | 412 | −2 | 24 | −41 | 394 |
| GDF SUEZ | 142 | 464 | 0 | 0 | −8 | 126 |
| IBERDROLA I | 340 | 722 | 28 | 331 | 18 | 292 |
| ING | −13 | 329 | 6 | 209 | −12 | 147 |
| INTESABCI | 1514 | 1096 | 0 | 0 | −20 | 658 |
| INDITEX | 547 | 702 | 3 | 400 | −10 | 198 |
| LVMH | 372 | 581 | 0 | 0 | −11 | 169 |
| MUNICH RE | 111 | 322 | −3 | 62 | −6 | 46 |
| LOREAL | 285 | 443 | 0 | 0 | −5 | 85 |
| PHILIPS ELECTR. | 113 | 417 | −6 | 96 | −8 | 105 |
| REPSOL | 611 | 809 | 40 | 254 | 27 | 437 |
| RWE ST | 38 | 450 | −2 | 299 | −42 | 372 |
| BANCO SAN CENTRAL HISPANO | 20 | 392 | −31 | 203 | 49 | 463 |
| SANOFI | 1 | 69 | 0 | 0 | −0 | 79 |

*Contd...*

|  | Order book imbalance | | Flow quantity | | Past return | |
| --- | --- | --- | --- | --- | --- | --- |
| Stock | $\overline{Gain}$ | $\sigma(Gain)$ | $\overline{Gain}$ | $\sigma(Gain)$ | $\overline{Gain}$ | $\sigma(Gain)$ |
| SAP AG | 2 | 120 | −4 | 137 | −30 | 207 |
| SAINT GOBAIN | 25 | 403 | −1 | 114 | −7 | 289 |
| SIEMENS AG | 2 | 74 | −2 | 89 | −3 | 141 |
| SCHNEIDER ELECTRIC SA | 16 | 317 | 0 | 0 | −14 | 195 |
| TELEFONICA | 656 | 663 | 6 | 139 | −7 | 183 |
| UNICREDIT SPA | 693 | 1159 | −5 | 173 | 19 | 628 |
| UNILEVER CERT | 1 | 56 | 0 | 0 | 0 | 0 |
| VIVENDI UNIVERSAL | 214 | 617 | 0 | 0 | −27 | 175 |
| VOLKSWAGEN | 171 | 545 | −7 | 115 | −45 | 246 |

**Table D.7** The quality of the 4-class prediction: The daily gain average and standard deviation for the 30-minute prediction (without trading costs)

|  | Order book imbalance | | Flow quantity | | Past return | |
| --- | --- | --- | --- | --- | --- | --- |
| Stock | $\overline{Gain}$ | $\sigma(Gain)$ | $\overline{Gain}$ | $\sigma(Gain)$ | $\overline{Gain}$ | $\sigma(Gain)$ |
| INTERBREW | −11 | 887 | −6 | 845 | −41 | 823 |
| AIR LIQUIDE | −57 | 669 | −17 | 633 | −21 | 624 |
| ALLIANZ | −14 | 762 | 69 | 689 | −41 | 729 |
| ASML Holding NV | −87 | 862 | 43 | 1075 | −29 | 897 |
| BASF AG | −20 | 807 | −3 | 781 | −67 | 722 |
| BAYER AG | 38 | 759 | −93 | 774 | −46 | 765 |
| BBVARGENTARIA | −16 | 1263 | −63 | 1138 | 16 | 1084 |
| BAY MOT WERKE | −25 | 783 | −23 | 923 | −13 | 901 |
| DANONE | −61 | 744 | 19 | 726 | −18 | 745 |
| BNP PARIBAS | −28 | 998 | −2 | 1179 | −9 | 1151 |
| CARREFOUR | 4 | 1108 | −135 | 1082 | −52 | 972 |
| CRH PLC IRLANDE | 75 | 962 | −105 | 1161 | −6 | 1117 |
| AXA | 12 | 1054 | 6 | 1055 | 49 | 1111 |
| DAIMLER CHRYSLER | 75 | 872 | −51 | 825 | 9 | 961 |
| DEUTSCHE BANK AG | 54 | 1054 | −89 | 1152 | −35 | 996 |
| VINCI | 110 | 761 | 80 | 742 | 100 | 743 |
| DEUTSCHE TELEKOM | 27 | 722 | 81 | 700 | −14 | 718 |

*Contd...*

|  | Order book imbalance | | Flow quantity | | Past return | |
|---|---|---|---|---|---|---|
| Stock | $\overline{Gain}$ | $\sigma(Gain)$ | $\overline{Gain}$ | $\sigma(Gain)$ | $\overline{Gain}$ | $\sigma(Gain)$ |
| ESSILOR INTERNATIONAL | 29 | 830 | 43 | 827 | 41 | 872 |
| ENEL | 27 | 991 | −40 | 971 | 55 | 959 |
| ENI | 7 | 628 | −18 | 651 | −16 | 645 |
| E.ON AG | −70 | 911 | −4 | 963 | 65 | 826 |
| TOTAL | 49 | 660 | 108 | 689 | 73 | 669 |
| GENERALI ASSIC | 18 | 1011 | 2 | 1094 | 11 | 1085 |
| SOCIETE GENERALE | 53 | 1413 | 67 | 1253 | −5 | 1335 |
| GDF SUEZ | 59 | 906 | −24 | 847 | 25 | 823 |
| IBERDROLA I | 3 | 1017 | −73 | 960 | 51 | 949 |
| ING | −21 | 1138 | 105 | 1205 | −80 | 1142 |
| INTESABCI | −128 | 1359 | −54 | 1329 | 85 | 1288 |
| INDITEX | −8 | 894 | −161 | 912 | 17 | 860 |
| LVMH | −36 | 831 | 15 | 725 | −26 | 675 |
| MUNICH RE | 29 | 641 | −25 | 688 | −7 | 727 |
| LOREAL | −19 | 671 | 31 | 755 | 15 | 727 |
| PHILIPS ELECTR. | −24 | 844 | 24 | 789 | −29 | 841 |
| REPSOL | −87 | 878 | −5 | 920 | 3 | 925 |
| RWE ST | 32 | 1132 | 61 | 1217 | 46 | 1140 |
| BANCO SAN CENTRAL HISPANO | 2 | 1150 | −60 | 1072 | 48 | 1090 |
| SANOFI | −29 | 810 | 25 | 856 | 7 | 794 |
| SAP AG | 4 | 683 | −52 | 709 | −15 | 682 |
| SAINT GOBAIN | −66 | 996 | 22 | 994 | −51 | 945 |
| SIEMENS AG | 127 | 771 | −35 | 802 | −59 | 725 |
| SCHNEIDER ELECTRIC SA | −31 | 896 | −79 | 837 | 8 | 838 |
| TELEFONICA | −12 | 759 | 42 | 918 | 111 | 912 |
| UNICREDIT SPA | 130 | 1529 | 58 | 1498 | 81 | 1357 |
| UNILEVER CERT | 5 | 543 | 31 | 546 | −26 | 508 |
| VIVENDI UNIVERSAL | 21 | 874 | −15 | 899 | 6 | 859 |
| VOLKSWAGEN | 71 | 929 | 120 | 994 | 75 | 1055 |

**Table D.8**  The quality of the binary prediction: The daily gain average and standard deviation for the 30-minute prediction (with 0.5 bp trading costs)

| Stock | Order book imbalance | | Flow quantity | | Past return | |
|---|---|---|---|---|---|---|
| | $\overline{Gain}$ | $\sigma(Gain)$ | $\overline{Gain}$ | $\sigma(Gain)$ | $\overline{Gain}$ | $\sigma(Gain)$ |
| INTERBREW | −55 | 887 | −51 | 845 | −84 | 824 |
| AIR LIQUIDE | −96 | 672 | −61 | 635 | −62 | 625 |
| ALLIANZ | −57 | 764 | 23 | 687 | −80 | 731 |
| ASML Holding NV | −132 | 863 | −6 | 1072 | −73 | 896 |
| BASF AG | −61 | 809 | −47 | 780 | −108 | 724 |
| BAYER AG | −7 | 758 | −136 | 777 | −84 | 767 |
| BBVARGENTARIA | −58 | 1265 | −108 | 1137 | −25 | 1082 |
| BAY MOT WERKE | −65 | 784 | −69 | 923 | −53 | 902 |
| DANONE | −101 | 743 | −25 | 726 | −60 | 742 |
| BNP PARIBAS | −71 | 997 | −46 | 1180 | −51 | 1149 |
| CARREFOUR | −39 | 1110 | −182 | 1085 | −94 | 972 |
| CRH PLC IRLANDE | 31 | 960 | −152 | 1163 | −48 | 1116 |
| AXA | −31 | 1052 | −37 | 1054 | 4 | 1109 |
| DAIMLER CHRYSLER | 36 | 874 | −93 | 825 | −31 | 961 |
| DEUTSCHE BANK AG | 9 | 1053 | −138 | 1151 | −77 | 997 |
| VINCI | 72 | 763 | 40 | 742 | 65 | 743 |
| DEUTSCHE TELEKOM | −12 | 722 | 36 | 702 | −53 | 720 |
| ESSILOR INTERNATIONAL | −9 | 830 | −1 | 828 | −2 | 869 |
| ENEL | −17 | 993 | −81 | 974 | 17 | 959 |
| ENI | −36 | 627 | −58 | 652 | −57 | 642 |
| E.ON AG | −106 | 911 | −45 | 965 | 22 | 824 |
| TOTAL | 10 | 661 | 66 | 690 | 34 | 666 |
| GENERALI ASSIC | −26 | 1011 | −44 | 1096 | −32 | 1087 |
| SOCIETE GENERALE | 10 | 1415 | 19 | 1252 | −51 | 1336 |
| GDF SUEZ | 14 | 905 | −70 | 847 | −16 | 818 |
| IBERDROLA I | −40 | 1016 | −117 | 962 | 5 | 947 |
| ING | −63 | 1137 | 58 | 1207 | −122 | 1144 |
| INTESABCI | −172 | 1359 | −97 | 1327 | 47 | 1290 |
| INDITEX | −48 | 896 | −204 | 913 | −22 | 859 |

*Contd...*

|  | Order book imbalance | | Flow quantity | | Past return | |
|---|---|---|---|---|---|---|
| Stock | $\overline{Gain}$ | $\sigma(Gain)$ | $\overline{Gain}$ | $\sigma(Gain)$ | $\overline{Gain}$ | $\sigma(Gain)$ |
| LVMH | −82 | 830 | −30 | 725 | −68 | 675 |
| MUNICH RE | −13 | 641 | −66 | 691 | −49 | 728 |
| LOREAL | −57 | 674 | −9 | 754 | −22 | 728 |
| PHILIPS ELECTR. | −65 | 845 | −23 | 788 | −71 | 839 |
| REPSOL | −128 | 877 | −52 | 920 | −41 | 920 |
| RWE ST | −7 | 1130 | 15 | 1218 | 5 | 1140 |
| BANCO SAN CENTRAL HISPANO | −37 | 1149 | −103 | 1073 | 6 | 1089 |
| SANOFI | −67 | 810 | −21 | 856 | −34 | 797 |
| SAP AG | −37 | 683 | −100 | 709 | −60 | 680 |
| SAINT GOBAIN | −105 | 997 | −23 | 995 | −93 | 946 |
| SIEMENS AG | 84 | 772 | −77 | 805 | −98 | 725 |
| SCHNEIDER ELECTRIC SA | −73 | 896 | −123 | 836 | −34 | 838 |
| TELEFONICA | −49 | 760 | −4 | 919 | 68 | 913 |
| UNICREDIT SPA | 84 | 1529 | 15 | 1499 | 40 | 1359 |
| UNILEVER CERT | −39 | 543 | −14 | 545 | −67 | 509 |
| VIVENDI UNIVERSAL | −24 | 874 | −61 | 900 | −37 | 856 |
| VOLKSWAGEN | 33 | 929 | 76 | 995 | 38 | 1058 |

Notice that the nans on the tables of the **Appendix 2** correspond to the cases where $|\widehat{Y}|$ is always lower than $\theta$ thus no positions are taken.

## D.3 Performances of the OLS Method

**Table D.9** The quality of the OLS prediction: The AUC and the accuracy per stock for the different horizons

|  | 1-min horizon | | 5-min horizon | | 30-min horizon | |
|---|---|---|---|---|---|---|
| Stock | AUC | Accuracy | AUC | Accuracy | AUC | Accuracy |
| INTERBREW | 0.54 | 0.54 | 0.50 | 0.50 | 0.50 | 0.50 |
| AIR LIQUIDE | 0.57 | 0.57 | 0.52 | 0.52 | 0.49 | 0.49 |
| ALLIANZ | 0.61 | 0.61 | 0.53 | 0.53 | 0.50 | 0.50 |
| ASML Holding NV | 0.55 | 0.55 | 0.51 | 0.51 | 0.51 | 0.51 |
| BASF AG | 0.54 | 0.54 | 0.52 | 0.52 | 0.50 | 0.50 |
| BAYER AG | 0.54 | 0.54 | 0.51 | 0.51 | 0.51 | 0.51 |

*Contd...*

|  | 1-min horizon | | 5-min horizon | | 30-min horizon | |
| --- | --- | --- | --- | --- | --- | --- |
| Stock | AUC | Accuracy | AUC | Accuracy | AUC | Accuracy |
| BBVARGENTARIA | 0.54 | 0.54 | 0.51 | 0.51 | 0.49 | 0.49 |
| BAY MOT WERKE | 0.55 | 0.55 | 0.51 | 0.51 | 0.49 | 0.49 |
| DANONE | 0.56 | 0.56 | 0.51 | 0.51 | 0.49 | 0.49 |
| BNP PARIBAS | 0.53 | 0.53 | 0.51 | 0.51 | 0.50 | 0.50 |
| CARREFOUR | 0.55 | 0.55 | 0.51 | 0.51 | 0.52 | 0.52 |
| CRH PLC IRLANDE | 0.62 | 0.62 | 0.56 | 0.56 | 0.52 | 0.52 |
| AXA | 0.55 | 0.55 | 0.51 | 0.51 | 0.50 | 0.50 |
| DAIMLER CHRYSLER | 0.54 | 0.54 | 0.51 | 0.51 | 0.50 | 0.50 |
| DEUTSCHE BANK AG | 0.53 | 0.53 | 0.51 | 0.51 | 0.51 | 0.51 |
| VINCI | 0.55 | 0.55 | 0.52 | 0.52 | 0.51 | 0.51 |
| DEUTSCHE TELEKOM | 0.56 | 0.56 | 0.52 | 0.52 | 0.50 | 0.51 |
| ESSILOR INTERNATIONAL | 0.56 | 0.56 | 0.51 | 0.51 | 0.51 | 0.51 |
| ENEL | 0.62 | 0.62 | 0.53 | 0.53 | 0.48 | 0.48 |
| ENI | 0.64 | 0.64 | 0.54 | 0.54 | 0.50 | 0.50 |
| E.ON AG | 0.57 | 0.57 | 0.52 | 0.52 | 0.48 | 0.48 |
| TOTAL | 0.54 | 0.54 | 0.51 | 0.51 | 0.50 | 0.50 |
| GENERALI ASSIC | 0.61 | 0.61 | 0.54 | 0.54 | 0.50 | 0.50 |
| SOCIETE GENERALE | 0.53 | 0.53 | 0.50 | 0.50 | 0.52 | 0.52 |
| GDF SUEZ | 0.56 | 0.56 | 0.51 | 0.51 | 0.50 | 0.50 |
| IBERDROLA I | 0.57 | 0.57 | 0.52 | 0.52 | 0.51 | 0.51 |
| ING | 0.53 | 0.53 | 0.51 | 0.51 | 0.49 | 0.49 |
| INTESABCI | 0.59 | 0.59 | 0.51 | 0.51 | 0.50 | 0.50 |
| INDITEX | 0.59 | 0.59 | 0.53 | 0.53 | 0.52 | 0.52 |
| LVMH | 0.59 | 0.59 | 0.52 | 0.52 | 0.52 | 0.52 |
| MUNICH RE | 0.58 | 0.58 | 0.53 | 0.53 | 0.50 | 0.50 |
| LOREAL | 0.60 | 0.60 | 0.52 | 0.52 | 0.51 | 0.51 |
| PHILIPS ELECTR. | 0.56 | 0.56 | 0.51 | 0.51 | 0.50 | 0.50 |
| REPSOL | 0.57 | 0.57 | 0.52 | 0.52 | 0.51 | 0.51 |
| RWE ST | 0.54 | 0.54 | 0.51 | 0.51 | 0.49 | 0.49 |
| BANCO SAN CENTRAL HISPANO | 0.54 | 0.54 | 0.51 | 0.51 | 0.49 | 0.49 |
| SANOFI | 0.54 | 0.54 | 0.51 | 0.51 | 0.49 | 0.49 |
| SAP AG | 0.54 | 0.54 | 0.51 | 0.51 | 0.51 | 0.51 |

*Contd...*

|  | 1-min horizon | | 5-min horizon | | 30-min horizon | |
|---|---|---|---|---|---|---|
| Stock | AUC | Accuracy | AUC | Accuracy | AUC | Accuracy |
| SAINT GOBAIN | 0.54 | 0.54 | 0.51 | 0.51 | 0.52 | 0.52 |
| SIEMENS AG | 0.54 | 0.54 | 0.51 | 0.51 | 0.50 | 0.50 |
| SCHNEIDER ELECTRIC SA | 0.54 | 0.54 | 0.52 | 0.52 | 0.51 | 0.51 |
| TELEFONICA | 0.59 | 0.59 | 0.52 | 0.52 | 0.50 | 0.50 |
| UNICREDIT SPA | 0.56 | 0.56 | 0.51 | 0.51 | 0.49 | 0.49 |
| UNILEVER CERT | 0.56 | 0.56 | 0.51 | 0.51 | 0.50 | 0.50 |
| VIVENDI UNIVERSAL | 0.57 | 0.57 | 0.51 | 0.51 | 0.51 | 0.51 |
| VOLKSWAGEN | 0.56 | 0.56 | 0.52 | 0.52 | 0.51 | 0.51 |

## D.4 Performances of the Ridge Method

**Table D.10** The quality of the Ridge HKB prediction: The AUC and the accuracy per stock for the different horizons

|  | 1-min horizon | | 5-min horizon | | 30-min horizon | |
|---|---|---|---|---|---|---|
| Stock | AUC | Accuracy | AUC | Accuracy | AUC | Accuracy |
| INTERBREW | 0.54 | 0.54 | 0.50 | 0.50 | 0.50 | 0.50 |
| AIR LIQUIDE | 0.57 | 0.57 | 0.52 | 0.52 | 0.50 | 0.50 |
| ALLIANZ | 0.61 | 0.61 | 0.53 | 0.53 | 0.49 | 0.49 |
| ASML Holding NV | 0.55 | 0.55 | 0.51 | 0.51 | 0.51 | 0.51 |
| BASF AG | 0.54 | 0.54 | 0.52 | 0.52 | 0.50 | 0.50 |
| BAYER AG | 0.54 | 0.54 | 0.51 | 0.51 | 0.50 | 0.50 |
| BBVARGENTARIA | 0.54 | 0.54 | 0.51 | 0.51 | 0.50 | 0.50 |
| BAY MOT WERKE | 0.55 | 0.55 | 0.51 | 0.51 | 0.50 | 0.50 |
| DANONE | 0.56 | 0.56 | 0.51 | 0.51 | 0.50 | 0.50 |
| BNP PARIBAS | 0.53 | 0.53 | 0.51 | 0.51 | 0.50 | 0.50 |
| CARREFOUR | 0.55 | 0.55 | 0.51 | 0.51 | 0.52 | 0.52 |
| CRH PLC IRLANDE | 0.62 | 0.62 | 0.56 | 0.56 | 0.52 | 0.52 |
| AXA | 0.56 | 0.55 | 0.51 | 0.51 | 0.50 | 0.50 |
| DAIMLER CHRYSLER | 0.54 | 0.54 | 0.51 | 0.51 | 0.50 | 0.50 |
| DEUTSCHE BANK AG | 0.53 | 0.53 | 0.51 | 0.51 | 0.51 | 0.51 |
| VINCI | 0.55 | 0.55 | 0.51 | 0.52 | 0.51 | 0.51 |
| DEUTSCHE TELEKOM | 0.56 | 0.56 | 0.52 | 0.52 | 0.51 | 0.52 |

*Contd...*

| Stock | 1-min horizon | | 5-min horizon | | 30-min horizon | |
| --- | --- | --- | --- | --- | --- | --- |
| | AUC | Accuracy | AUC | Accuracy | AUC | Accuracy |
| ESSILOR INTERNATIONAL | 0.56 | 0.56 | 0.51 | 0.51 | 0.51 | 0.51 |
| MUNICH RE | 0.59 | 0.59 | 0.53 | 0.53 | 0.50 | 0.50 |
| ENEL | 0.62 | 0.62 | 0.53 | 0.53 | 0.48 | 0.48 |
| ENI | 0.65 | 0.65 | 0.54 | 0.54 | 0.50 | 0.50 |
| E.ON AG | 0.57 | 0.57 | 0.52 | 0.52 | 0.48 | 0.48 |
| TOTAL | 0.54 | 0.54 | 0.51 | 0.51 | 0.50 | 0.50 |
| GENERALI ASSIC | 0.62 | 0.62 | 0.54 | 0.54 | 0.51 | 0.51 |
| SOCIETE GENERALE | 0.53 | 0.53 | 0.50 | 0.50 | 0.53 | 0.52 |
| GDF SUEZ | 0.57 | 0.57 | 0.52 | 0.52 | 0.50 | 0.50 |
| IBERDROLA I | 0.57 | 0.57 | 0.53 | 0.53 | 0.52 | 0.52 |
| ING | 0.53 | 0.53 | 0.51 | 0.50 | 0.50 | 0.50 |
| INTESABCI | 0.60 | 0.60 | 0.52 | 0.52 | 0.50 | 0.50 |
| INDITEX | 0.59 | 0.59 | 0.53 | 0.53 | 0.52 | 0.52 |
| LVMH | 0.59 | 0.59 | 0.52 | 0.52 | 0.50 | 0.50 |
| LOREAL | 0.60 | 0.60 | 0.52 | 0.52 | 0.51 | 0.51 |
| PHILIPS ELECTR. | 0.56 | 0.56 | 0.51 | 0.51 | 0.49 | 0.49 |
| REPSOL | 0.58 | 0.58 | 0.52 | 0.52 | 0.52 | 0.52 |
| RWE ST | 0.54 | 0.54 | 0.51 | 0.51 | 0.50 | 0.50 |
| BANCO SAN CENTRAL HISPANO | 0.54 | 0.54 | 0.51 | 0.51 | 0.50 | 0.50 |
| SANOFI | 0.54 | 0.54 | 0.51 | 0.51 | 0.51 | 0.51 |
| SAP AG | 0.55 | 0.55 | 0.51 | 0.51 | 0.51 | 0.51 |
| SAINT GOBAIN | 0.54 | 0.54 | 0.51 | 0.51 | 0.52 | 0.52 |
| SIEMENS AG | 0.54 | 0.54 | 0.51 | 0.51 | 0.51 | 0.51 |
| SCHNEIDER ELECTRIC SA | 0.55 | 0.55 | 0.52 | 0.52 | 0.50 | 0.50 |
| TELEFONICA | 0.59 | 0.59 | 0.52 | 0.52 | 0.51 | 0.51 |
| UNICREDIT SPA | 0.57 | 0.57 | 0.51 | 0.51 | 0.49 | 0.49 |
| UNILEVER CERT | 0.56 | 0.56 | 0.51 | 0.51 | 0.49 | 0.49 |
| VIVENDI UNIVERSAL | 0.57 | 0.57 | 0.51 | 0.51 | 0.51 | 0.51 |
| VOLKSWAGEN | 0.57 | 0.57 | 0.52 | 0.52 | 0.51 | 0.51 |

**Table D.11**   The quality of the Ridge LW prediction: The AUC and the accuracy per stock for the different horizons

| Stock | 1-min horizon | | 5-min horizon | | 30-min horizon | |
|---|---|---|---|---|---|---|
| | AUC | Accuracy | AUC | Accuracy | AUC | Accuracy |
| INTERBREW | 0.55 | 0.55 | 0.52 | 0.52 | 0.50 | 0.50 |
| AIR LIQUIDE | 0.57 | 0.57 | 0.53 | 0.53 | 0.49 | 0.49 |
| ALLIANZ | 0.61 | 0.61 | 0.54 | 0.54 | 0.50 | 0.50 |
| ASML Holding NV | 0.56 | 0.56 | 0.52 | 0.52 | 0.52 | 0.52 |
| BASF AG | 0.54 | 0.54 | 0.52 | 0.52 | 0.50 | 0.50 |
| BAYER AG | 0.55 | 0.55 | 0.51 | 0.51 | 0.50 | 0.50 |
| BBVARGENTARIA | 0.54 | 0.54 | 0.51 | 0.51 | 0.50 | 0.50 |
| BAY MOT WERKE | 0.55 | 0.55 | 0.51 | 0.51 | 0.50 | 0.50 |
| DANONE | 0.56 | 0.56 | 0.51 | 0.51 | 0.49 | 0.49 |
| BNP PARIBAS | 0.54 | 0.54 | 0.52 | 0.52 | 0.50 | 0.50 |
| CARREFOUR | 0.55 | 0.55 | 0.51 | 0.51 | 0.51 | 0.51 |
| CRH PLC IRLANDE | 0.62 | 0.62 | 0.57 | 0.57 | 0.51 | 0.51 |
| AXA | 0.56 | 0.56 | 0.51 | 0.51 | 0.51 | 0.51 |
| DAIMLER CHRYSLER | 0.54 | 0.54 | 0.52 | 0.52 | 0.51 | 0.51 |
| DEUTSCHE BANK AG | 0.53 | 0.53 | 0.51 | 0.51 | 0.52 | 0.52 |
| VINCI | 0.56 | 0.56 | 0.52 | 0.53 | 0.51 | 0.52 |
| DEUTSCHE TELEKOM | 0.57 | 0.57 | 0.52 | 0.52 | 0.52 | 0.52 |
| ESSILOR INTERNATIONAL | 0.56 | 0.56 | 0.51 | 0.51 | 0.50 | 0.50 |
| ENEL | 0.63 | 0.63 | 0.54 | 0.54 | 0.50 | 0.50 |
| ENI | 0.65 | 0.65 | 0.55 | 0.55 | 0.50 | 0.50 |
| E.ON AG | 0.58 | 0.58 | 0.52 | 0.52 | 0.50 | 0.51 |
| TOTAL | 0.54 | 0.54 | 0.52 | 0.52 | 0.51 | 0.51 |
| GENERALI ASSIC | 0.62 | 0.62 | 0.55 | 0.55 | 0.49 | 0.49 |
| SOCIETE GENERALE | 0.53 | 0.53 | 0.50 | 0.50 | 0.52 | 0.52 |
| GDF SUEZ | 0.57 | 0.57 | 0.52 | 0.52 | 0.50 | 0.50 |
| IBERDROLA I | 0.57 | 0.57 | 0.53 | 0.53 | 0.52 | 0.52 |
| ING | 0.53 | 0.53 | 0.51 | 0.51 | 0.50 | 0.50 |
| INTESABCI | 0.60 | 0.60 | 0.53 | 0.53 | 0.48 | 0.48 |
| INDITEX | 0.60 | 0.60 | 0.54 | 0.54 | 0.51 | 0.51 |

*Contd...*

|  | 1-min horizon | | 5-min horizon | | 30-min horizon | |
|---|---|---|---|---|---|---|
| Stock | AUC | Accuracy | AUC | Accuracy | AUC | Accuracy |
| LVMH | 0.59 | 0.59 | 0.52 | 0.52 | 0.50 | 0.50 |
| MUNICH RE | 0.59 | 0.59 | 0.54 | 0.54 | 0.50 | 0.50 |
| LOREAL | 0.60 | 0.60 | 0.53 | 0.53 | 0.51 | 0.51 |
| PHILIPS ELECTR. | 0.57 | 0.57 | 0.52 | 0.52 | 0.50 | 0.50 |
| REPSOL | 0.58 | 0.58 | 0.53 | 0.53 | 0.51 | 0.51 |
| RWE ST | 0.55 | 0.55 | 0.51 | 0.51 | 0.49 | 0.49 |
| BANCO SAN CENTRAL HISPANO | 0.54 | 0.54 | 0.52 | 0.52 | 0.51 | 0.51 |
| SANOFI | 0.55 | 0.55 | 0.51 | 0.51 | 0.50 | 0.50 |
| SAP AG | 0.55 | 0.55 | 0.51 | 0.51 | 0.51 | 0.51 |
| SAINT GOBAIN | 0.55 | 0.55 | 0.51 | 0.51 | 0.52 | 0.52 |
| SIEMENS AG | 0.55 | 0.55 | 0.52 | 0.52 | 0.51 | 0.52 |
| SCHNEIDER ELECTRIC SA | 0.55 | 0.55 | 0.52 | 0.52 | 0.50 | 0.50 |
| TELEFONICA | 0.60 | 0.60 | 0.53 | 0.53 | 0.51 | 0.51 |
| UNICREDIT SPA | 0.57 | 0.57 | 0.52 | 0.52 | 0.49 | 0.49 |
| UNILEVER CERT | 0.57 | 0.57 | 0.51 | 0.51 | 0.51 | 0.51 |
| VIVENDI UNIVERSAL | 0.58 | 0.58 | 0.52 | 0.52 | 0.51 | 0.51 |
| VOLKSWAGEN | 0.57 | 0.57 | 0.52 | 0.52 | 0.50 | 0.50 |

## D.5 Performances of the LASSO Method

**Table D.12** The quality of the LASSO prediction: The AUC and the accuracy per stock for the different horizons

|  | 1-min horizon | | 5-min horizon | | 30-min horizon | |
|---|---|---|---|---|---|---|
| Stock | AUC | Accuracy | AUC | Accuracy | AUC | Accuracy |
| INTERBREW | 0.54 | 0.54 | 0.51 | 0.51 | 0.50 | 0.50 |
| AIR LIQUIDE | 0.58 | 0.58 | 0.52 | 0.52 | 0.49 | 0.49 |
| ALLIANZ | 0.61 | 0.61 | 0.54 | 0.54 | 0.52 | 0.52 |
| ASML Holding NV | 0.56 | 0.56 | 0.52 | 0.52 | 0.51 | 0.51 |
| BASF AG | 0.53 | 0.53 | 0.51 | 0.51 | 0.51 | 0.51 |
| BAYER AG | 0.54 | 0.54 | 0.51 | 0.51 | 0.50 | 0.50 |
| BBVARGENTARIA | 0.54 | 0.54 | 0.51 | 0.51 | 0.49 | 0.49 |

Contd...

|  | 1-min horizon | | 5-min horizon | | 30-min horizon | |
| --- | --- | --- | --- | --- | --- | --- |
| Stock | AUC | Accuracy | AUC | Accuracy | AUC | Accuracy |
| BAY MOT WERKE | 0.55 | 0.55 | 0.51 | 0.51 | 0.49 | 0.49 |
| DANONE | 0.56 | 0.56 | 0.51 | 0.51 | 0.50 | 0.50 |
| BNP PARIBAS | 0.54 | 0.54 | 0.51 | 0.51 | 0.49 | 0.49 |
| CARREFOUR | 0.55 | 0.55 | 0.51 | 0.51 | 0.50 | 0.50 |
| CRH PLC IRLANDE | 0.62 | 0.62 | 0.56 | 0.56 | 0.52 | 0.52 |
| AXA | 0.55 | 0.55 | 0.51 | 0.51 | 0.49 | 0.49 |
| DAIMLER CHRYSLER | 0.53 | 0.53 | 0.52 | 0.52 | 0.50 | 0.50 |
| DEUTSCHE BANK AG | 0.53 | 0.53 | 0.51 | 0.51 | 0.51 | 0.51 |
| VINCI | 0.55 | 0.55 | 0.52 | 0.53 | 0.52 | 0.52 |
| DEUTSCHE TELEKOM | 0.58 | 0.58 | 0.52 | 0.52 | 0.52 | 0.52 |
| ESSILOR INTERNATIONAL | 0.56 | 0.56 | 0.51 | 0.51 | 0.50 | 0.50 |
| ENEL | 0.62 | 0.62 | 0.53 | 0.53 | 0.50 | 0.50 |
| ENI | 0.64 | 0.64 | 0.55 | 0.55 | 0.49 | 0.49 |
| E.ON AG | 0.57 | 0.57 | 0.52 | 0.52 | 0.49 | 0.50 |
| TOTAL | 0.54 | 0.54 | 0.52 | 0.52 | 0.51 | 0.51 |
| GENERALI ASSIC | 0.62 | 0.62 | 0.54 | 0.54 | 0.51 | 0.51 |
| SOCIETE GENERALE | 0.53 | 0.53 | 0.50 | 0.50 | 0.52 | 0.52 |
| GDF SUEZ | 0.56 | 0.56 | 0.52 | 0.52 | 0.51 | 0.51 |
| IBERDROLA I | 0.56 | 0.56 | 0.53 | 0.53 | 0.53 | 0.53 |
| ING | 0.52 | 0.52 | 0.51 | 0.51 | 0.50 | 0.50 |
| INTESABCI | 0.60 | 0.60 | 0.53 | 0.53 | 0.50 | 0.50 |
| INDITEX | 0.59 | 0.59 | 0.53 | 0.53 | 0.52 | 0.52 |
| LVMH | 0.59 | 0.59 | 0.52 | 0.52 | 0.51 | 0.51 |
| MUNICH RE | 0.58 | 0.58 | 0.54 | 0.54 | 0.50 | 0.50 |
| LOREAL | 0.60 | 0.60 | 0.53 | 0.53 | 0.50 | 0.50 |
| PHILIPS ELECTR. | 0.56 | 0.56 | 0.52 | 0.52 | 0.50 | 0.50 |
| REPSOL | 0.57 | 0.57 | 0.52 | 0.52 | 0.51 | 0.51 |
| RWE ST | 0.54 | 0.54 | 0.51 | 0.51 | 0.50 | 0.50 |
| BANCO SAN CENTRAL HISPANO | 0.54 | 0.54 | 0.52 | 0.52 | 0.50 | 0.50 |
| SANOFI | 0.54 | 0.54 | 0.51 | 0.51 | 0.50 | 0.50 |
| SAP AG | 0.53 | 0.53 | 0.52 | 0.52 | 0.50 | 0.50 |

*Contd...*

|  | 1-min horizon | | 5-min horizon | | 30-min horizon | |
|---|---|---|---|---|---|---|
| Stock | AUC | Accuracy | AUC | Accuracy | AUC | Accuracy |
| SAINT GOBAIN | 0.54 | 0.54 | 0.51 | 0.51 | 0.52 | 0.52 |
| SIEMENS AG | 0.54 | 0.54 | 0.51 | 0.51 | 0.50 | 0.50 |
| SCHNEIDER ELECTRIC SA | 0.54 | 0.54 | 0.51 | 0.51 | 0.49 | 0.49 |
| TELEFONICA | 0.59 | 0.59 | 0.53 | 0.53 | 0.51 | 0.51 |
| UNICREDIT SPA | 0.57 | 0.57 | 0.52 | 0.52 | 0.48 | 0.48 |
| UNILEVER CERT | 0.57 | 0.57 | 0.51 | 0.51 | 0.51 | 0.51 |
| VIVENDI UNIVERSAL | 0.57 | 0.57 | 0.52 | 0.52 | 0.52 | 0.52 |
| VOLKSWAGEN | 0.56 | 0.56 | 0.52 | 0.52 | 0.49 | 0.49 |

# Bibliography

Abergel, F., and A. Jedidi. 2013. 'A Mathematical Approach to Order Book Modelling'. *International Journal of Theoretical and Applied Finance* 16, no. 5: 1350025-1-1350025-40.

———. 'Long-Time Behavior of a Hawkes Process-Based Limit Order Book'. *SIAM Journal on Financial Mathematics* 6: 1026–1043.

Abergel, F., and M. Politi. 2013. 'Optimizing a Basket Against the Efficient Market Hypothesis'. *Quantitative Finance* 13, no. 1: 13–23.

Abergel, F., C. A. Lehalle, and M. Rosenbaum. 2014. 'Understanding the Stakes of High-Frequency Trading'. *Journal of Trading* 9, no. 4: 49–73.

Alfi, V., M. Cristelli, L. Pietronero, and A. Zaccaria. 2009a. 'Minimal Agent Based Model for Financial Markets I : Origin and Self-organization of Stylized Facts'. *The European Physical Journal B* 67: 385–397.

———. 2009b. 'Minimal Agent Based Model for Financial Markets II'. *The European Physical Journal B* 67: 399–417.

Alfonsi, A., and P. Blanc. 2015. 'Dynamic Optimal Execution in a Mixed-market-impact Hawkes Price Model'. Accessed September 29, 2015. http://arxiv.org/abs/1404.0648

Anane, M., and F. Abergel. 2015. 'Empirical Evidence of Market Inefficiency: Predicting Single-Stock Returns'. In *Econophysics and Data Driven Modelling of Market Dynamics.*, edited by F. Abergel, H. Aoyama, B. K. Chakrabarti, A. Chakraborti, and A. Ghosh, 3-66. Cham (ZG), Switzerland: Springer International Publishing.

Anane, M., F. Abergel, and X. F. Lu. 2015. 'Mathematical modeling of the order book: new approach for predicting single-stock returns'. Working paper.

Bacry, E., K. Dayri, and J. F. Muzy. 2012. 'Non-parametric kernel estimation for symmetric Hawkes processes. Application to high frequency financial data'. *The European Physics Journal B* 85: 157.

Bacry, E., S. Delattre, M. Hoffmann, and J. F. Muzy. 2013a. 'Modelling microstructure noise with mutually exciting point processes'. *Quantitative Finance* 13, no. 1: 65–77.

———. 2013b. 'Scaling limits for Hawkes processes and application to financial statistics'. *Stochastic Processes and Applications* 123: 2475–2499.

Bacry, E., T. Jaisson, and J. F. Muzy. 2014. 'Estimation of slowly decreasing Hawkes kernels: Application to high frequency order book modelling'. Accessed September 29, 2015. http://arxiv.org/abs/1412.7096.

Bak, P., M. Paczuski, and M. Shubik. 1997. 'Price variations in a stock market with many agents'. *Physica A: Statistical and Theoretical Physics* 246: 430–453.

Bhattacharya, R. N. 1982. 'On the Functional Central Limit Theorem and the Law of the Iterated Logarithm for Markov Processes'. *Zeitschrift für Wahrscheinlichkeitstheorie* 60: 185–201.

Biais, B., P. Hillion, and C. Spatt. 1995. 'An Empirical Analysis of the Limit Order Book and the Order Flow in the Paris Bourse'. *The Journal of Finance* 50: 1655–1689.

Biais, B., T. Foucault, and P. Hillion. 1997. *'Microstructure des marches financiers: Institutions, modeles et tests empiriques'*. Paris: Presses Universitaires de France.

Bouchaud, J. P., M. Mezard, and M. Potters. 2002. 'Statistical properties of stock order books: Empirical results and models'. *Quantitative Finance* 2, no. 4: 251–256.

Bouchaud, J. P., and M. Potters. 2004. *Theory of Financial Risk and Derivative Pricing: From Statistical Physics to Risk Management.* Second Edition Cambridge: Cambridge University Press.

Bouchaud, J. P., Y. Gefen, M. Potters, and M. Wyart. 2004. 'Fluctuations and response in financial markets: The subtle nature of random price changes'. *Quantitative Finance* 4, no. 2: 176–190.

Bouchaud, J. P., J. D. Farmer, and F. Lillo. 2009. 'How Markets Slowly Digest Changes in supply and Demand'. In *Handbook of Financial Markets: Dynamics and Evolution*, T. Hens, and K. R. Schenk-Hoppé, 57–156 San Diego: Elsevier

Bowsher, C. G. 2007. 'Modelling security market events in continuous time: Intensity based, multivariate point process models'. *Journal of Econometrics* 141: 876–912.

Brémaud, P. 1981. *Point Processes and Queues: Martingale Dynamics.* New York: Springer.

———. 1999. *Markov Chains: Gibbs Fields, Monte Carlo Simulation, and Queues.* New York: Springer.

Brock, W. A., and C. H. Hommes. 1998. 'Heterogeneous beliefs and routes to chaos in a simple asset pricing model'. *Journal of Economic Dynamics and Control* 22: 1235–1274.

Cattiaux, P., D. Chafaï, and A. Guillin. 2012. 'Central limit theorems for additive functionals of ergodic Markov diffusions processes'. *Latin American Journal of Probability and Mathematical Statistics* 9: 337–382.

Chakraborti, A., I. Muni Toke, M. Patriarca, and F. Abergel. 2011a. 'Econophysics review: I. Empirical facts'. *Quantitative Finance* 11, no. 7: 991–1012.

———. 2011b. 'Econophysics review: II. Agent-based models'. *Quantitative Finance* 11, no. 7: 1013–1041.

Challet, D., and R. Stinchcombe. 2001. 'Analyzing and modeling 1 + 1d markets'. *Physica A: Statistical Mechanics and its Applications* 300: 285–299.

———. 2003. 'Non-constant rates and over-diffusive prices in a simple model of limit order markets'. *Quantitative Finance* 3: 155–162.

Chaudhry, M. L., and J. G. C. Templeton. 1983. *'A first course in bulk queues'*. New York: Wiley.

Cheney, E., and D. Kincaid. 2008. *Numerical Mathematics and Computing*. Boston: Cengage Learning.

Chiarella, C., and G. Iori. 2002. 'A simulation analysis of the microstructure of double auction markets'. *Quantitative Finance* 2: 346–353.

Clark, P. K. 1973. 'A Subordinated Stochastic Process Model with Finite Variance for Speculative Prices'. *Econometrica* 41: 135–55.

Cliff, D., and J. Bruten. 1997. *'Zero is not enough: On the lower limit of agent intelligence for continuous double auction markets'*. Tech. rept. HPL-97-141, Bristol, UK: Hewlett- Packard Laboratories.

Clinet, S. 2015. *Some ergodicity results for Limit Order Books*.

Cohen, M. H. 1963a. *'Reflections on the Special Study of the Securities Markets'*. Accessed September 29, 2015.
https://www.sec.gov/news/speech/1963/051063cohen.pdf

———. 1963b. *'Report of the Special Study of the Securities Markets of the Securities and Exchanges Commission'*. Accessed September 29, 2015.
http://3197d6d14b5f19f2f440−5e13d29c4c016cf96cbbfd197c579b45.r81.cf1.rackcdn.com/collection/papers/1960/1963_SSMkt_Chapter_01_1.pdf

Cont, R. 2007. 'Volatility Clustering in Financial Markets: Empirical Facts and Agent-Based Models'. In *Long Memory in Economics*, edited by G. Teyssiere, and A. P. Kirman, 289-309. Heidelberg: Springer.

Cont, R., and J. P. Bouchaud. 2000. 'Herd Behavior and Aggregate Fluctuations in Financial Markets'. *Macroeconomic Dynamics* 4: 170–196.

Cont, R., and A. de Larrard. 2012. 'Order Book Dynamics in Liquid Markets: Limit Theorems and Diffusion Approximations'. Working paper.

Cont, R., S. Stoikov, and R. Talreja. 2010. A Stochastic Model for Order Book Dynamics. *Operations Research* 58: 549–563.

Cont, R., A. Kukanov, and S. Stoikov. 2014. 'The Price Impact of Order Book Events'. *Journal of financial econometrics* 12: 47–88.

Da Fonseca, J., and R. Zaatour. 2014a. 'Clustering and Mean Reversion in a Hawkes Microstructure Model'. *Journal of Futures Markets* 35: 813–838.

———. 2014b. 'Hawkes Process: Fast Calibration, Application to Trade Clustering, and Diffusive Limit'. *Journal of Futures Markets* 34: 548–579.

Daley, D. J., and D. Vere-Jones. 2003. *An Introduction to the Theory of Point Processes. Vol. I: Elementary Theory and Methods.*. New York: Springer-Verlag.

———. 2008. *An Introduction to the Theory of Point Processes. Vol. II: General Theory and Structure*. New York: Springer-Verlag.

de Haan, L., S. Resnick, and H. Drees. 2000. 'How to make a Hill plot'. *The Annals of Statistics* 28: 254–274.

Dremin, I. M., and A.V. Leonidov. 2005. 'On distribution of number of trades in different time windows in the stock market'. *Physica A: Statistical Mechanics and its Applications*, 353: 388–402.

Duflo, M. 1990. *Méthodes récursives aléatoires*. Paris: Masson.

Engle, R. F. 2000. 'The Econometrics of Ultra-High-Frequency Data'. *Econometrica* 68: 1–22.

Engle, R. F., and J. R. Russell. 1997. 'Forecasting the frequency of changes in quoted foreign exchange prices with the autoregressive conditional duration model'. *Journal of Empirical Finance* 4: 187–212.

Erdos, P., and A. Renyi. 1960. 'On the Evolution of Random Graphs'. *Publications of the Mathematical Institute of the Hungarian Academy of Sciences* 5: 17–61.

Ethier, S. N., and T. G. Kurtz. 2005. *Markov Processes: Characterization and Convergence*. Hoboken: Wiley.

Farmer, J. D., A. Gerig, F. Lillo, and S. Mike. 2006. 'Market efficiency and the longmemory of supply and demand: Is price impact variable and permanent or fixed and temporary?' *Quantitative Finance* 6: 107–112.

Feller, W. 1968. *Introduction to the Theory of Probability and its Applications* Vol. 2. New York: Wiley.

Filimonov, V., and D. Sornette. 2015. 'Apparent criticality and calibration issues in the Hawkes self-excited point process model: Application to high-frequency financial data'. *Quantitative Finance* 15: 1293–1314.

Gao, X., J. G. Dai, A. B. Dieker, and S. J. Deng. 2014. 'Hydrodynamic Limit of Order Book Dynamics'. Working paper. Accessed September 29, 2015.
http://papers.ssrn.com/sol3/papers.cfm?abstract_id=2530306

Garman, M. B. 1976. 'Market microstructure'. *Journal of Financial Economics* 3: 257–275.

Gatheral, J., and R. Oomen. 2010. 'Zero-Intelligence Realized Variance Estimation'. *Finance and Stochastics* 14: 249–283.

Gatheral, J., T. Jaisson, T., and M. Rosenbaum. 2015. 'Volatility is rough'. Accessed September 29, 2015.
https://www.researchgate.net/publication/266856321_Volatility_is_rough

Glosten, L. R. 1994. 'Is the Electronic Open Limit Order Book Inevitable?' *The Journal of Finance* 49: 1127–1161.

Glynn, P. W., and S. P. Meyn. 1996. 'A Liapounov bound for solutions of the Poisson equation'. *Annals of Probability* 24: 916–931.

Gode, D. K., and S. Sunder. 1993. 'Allocative Efficiency of Markets with Zero-Intelligence Traders: Market as a Partial Substitute for Individual Rationality'. *The Journal of Political Economy* 101: 119–137.

Gopikrishnan, P., V. Plerou, X. Gabaix, and H. E. Stanley. 2000. 'Statistical properties of share volume traded in financial markets'. *Physical Review E* 62: 4493–4496.

Gu, G. F., and W. X. Zhou. 2009. 'Emergence of long memory in stock volatility from a modified MikeFarmer model'. *Europhysics Letters* 86: 48002.

Gu, G. F., W. Chen, and W. X. Zhou. 2008. 'Empirical shape function of limit-order books in the Chinese stock market'. *Physica A: Statistical Mechanics and its Applications* 387: 5182–5188.

Guo, Xin, Ruan, Zhao, and Zhu, Lingjiong. 2015. 'Dynamics of Order Positions and Related Queues in a Limit Order Book'. Accessed September 29, 2015.
http://arxiv.org/pdf/1505.04810.pdf

Hakansson, N. H., A. Beja, and J. Kale. 1985. 'On the Feasibility of Automated Market Making by a Programmed Specialist'. *Journal of Finance* 40: 1–20.

Hardiman, S. J., N. Bercot, and J. P. Bouchaud. 2013. 'Critical reflexivity in financial markets: A Hawkes process analysis'. *The European Physical Journal B* 86: 1–9.

Hasbrouck, J. 2007. *Empirical Market Microstructure: The Institutions, Economics, and Econometrics of Securities Trading.* New York: Oxford University Press.

Hastie, T., R. Tibshirani, and J. Friedman. 2011. *The Elements of Statistical Learning.* New York: Springer.

Hautsch, N. 2004. *Modelling Irregularly Spaced Financial Data.* Heidelberg: Springer.

Hawkes, A. G. 1971. 'Spectra of some self-exciting and mutually exciting point processes'. *Biometrika* 58: 83–90.

Hawkes, A. G., and D. Oakes. 1974. 'A Cluster Process Representation of a Self-Exciting Process'. *Journal of Applied Probability* 11: 493–503.

Hill, B. M. 1975. 'A Simple General Approach to Inference About the Tail of a Distribution'. *The Annals of Statistics* 3: 1163–1174.

Hoerl, A. E., and R. W. Kennard. 1970. 'Ridge Regression: Biased Estimation for Nonorthogonal Problems'. *Technometrics* 12: 55–67.

Hoerl, A. E., R. W. Kennard, and K. F. Baldwin. 1975. 'Ridge regression: Some simulations'. *Communications in Statistics-Theory and Methods* 4: 105–123.

Horst, U., and M. Paulsen. 2015. 'A law of large numbers for limit order books'. Accessed September 29, 2015.
http://arxiv.org/abs/1501.00843

Huang, W., and M. Rosenbaum. 2015. 'Ergodicity and diffusivity of Markovian order book models: A general framework'. Accessed September 29, 2015.
http://arxiv.org/abs/1505.04936

Huang, W., C. A. Lehalle, and M. Rosenbaum. 2015. 'Simulating and analyzing order book data: The queue-reactive model'. *Journal of the American Statistical Association* 110: 107–122.

Huth, N., and F. Abergel. 2012. 'The times change: Multivariate subordination. Empirical facts'. *Quantitative Finance* 12: 1–10.

Ivanov, P. C., A. Yuen, B. Podobnik, and Y. Lee. 2004. 'Common scaling patterns in intertrade times of U. S. stocks'. *Physical Review E* 69: 056107.

Jacod, J., and A. N. Shyriaev. 2003. *Limit Theorems for Stochastic Processes*. Heidelberg: Springer.

Kirman, A. P. 1983. 'On mistaken beliefs and resultant equilibria. In *Individual Forecasting and Aggregate Outcomes: Rational Expectations Examined*. Cambridge: Cambridge University Press.

Kirman, A. P. 1992. 'Whom or What Does the Representative Individual Represent?' *The Journal of Economic Perspectives* 6: 117–136.

———. 1993. 'Ants, Rationality, and Recruitment' *The Quarterly Journal of Economics* 108: 137–156.

———. 2002. 'Reflections on Interaction and Markets'. *Quantitative Finance* 2: 322–326.

Kyle, A. S. 1985. 'Continuous Auctions and Insider Trading'. *Econometrica* 53: 1315–1335.

Lallouache, M., and D. Challet. 2015. 'The limits of statistical significance of Hawkes processes fitted to financial data'. Accessed September 29, 2015. http://arxiv.org/pdf/1406.3967v1.pdf

Large, J. H. 2007. 'Measuring the Resiliency of an Electronic Limit Order Book'. *Journal of Financial Markets* 10: 1–25.

Lawless, J. F., and P. Wang. 1976. 'A simulation study of ridge and other regression estimators'. *Communications in Statistics - Theory and Methods* 5: 307–323.

LeBaron, B. 2006a. 'Agent-based Computational Finance'. In *Handbook of Computational Economics*. Vol. 2., edited by Leigh Tesfatsion and Kenneth L. Judd, 1187–1233. Amsterdam: North-Holland.

———. 2006b. 'Agent-Based Financial Markets: Matching Stylized Facts with Style'. In *Post Walrasian Macroeconomics*, edited by David Colander, 221–35. Cambridge: Cambridge University Press.

Lewis, P. A. W, and G. S. Shedler. 1979. 'Simulation of Nonhomogeneous Poisson Processes by Thinning'. *Naval Research Logistics Quarterly* 26: 403–413.

Lillo, F., and J. D. Farmer. 2004. 'The Long Memory of the Efficient Market' *Studies in Nonlinear Dynamics & Econometrics* 8: 1558–3708.

Lux, T., and M. Marchesi. 2000. 'Volatility Clustering in Financial Markets: A Microsimulation of Interacting Agents'. *International Journal of Theoretical and Applied Finance* 3: 675–702.

Mandelbrot, B. 1963. 'The Variation of Certain Speculative Prices'. *Journal of Business* 36: 394–419.

Maruyama, G., and H. Tanaka. 1959. 'Ergodic property of $N$-dimensional recurrent Markov processes'. *Memoirs of the Faculty of Science, Kyushu University. Series A - Mathematics* 13: 157–172.

Maslov, S. 2000. 'Simple model of a limit order-driven market'. *Physica A: Statistical Mechanics and its Applications* 278: 571–578.

Maslov, S., and M. Mills 2001. 'Price fluctuations from the order book perspective – empirical facts and a simple model'. *Physica A: Statistical Mechanics and its Applications* 299: 234–246.

Massoulié, L. 1998. 'Stability results for a general class of interacting point processes dynamics, and applications'. *Stochastic Processes and their Applications* 75: 1–30.

Meyn, S., and R. L. Tweedie. 1993. 'Stability of Markovian processes III: Foster-Lyapunov criteria for continuous-time processes'. *Advances in Applied Probability* 25: 518–548.

———. 2009. *Markov Chains and Stochastic Stability*. New York: Cambridge University Press.

Mike, S., and J. D. Farmer. 2008. 'An Empirical Behavioral Model of Liquidity and Volatility'. *Journal of Economic Dynamics and Control* 32: 200–234.

Muni Toke, I. 2011. 'Market Making in an Order Book Model and Its Impact on the Spread'. In *Econophysics of Order-driven Markets*, edited by F. Abergel, B. K. Chakrabarti, A. Chakraborti, and M. Mitra, 49–64. Milan: Springer.

Muni Toke, I. 2015. 'The order book as a queueing system: average depth and influence of the size of limit orders'. *Quantitative Finance* 15: 795–808.

Muni Toke, I., and F. Pomponio. 2012. 'Modelling trade-throughs in a limited order book using Hawkes processes'. *Economics: The Open-Access, Open-Assessment E-Journal* 6.

Murphy, J. 1999. *Technical Analysis of the Financial Markets: A Comprehensive Guide to Trading Methods and Applications*. New York: New York Institute of Finance.

Naes, R., and J. A. Skjeltorp. 2006. 'Order Book Characteristics and the Volume-Volatility Relation: Empirical Evidence from a Limit Order Market'. *Journal of Financial Markets* 9: 408–432.

O'Hara, M. 1997. *Market Microstructure Theory*. Cambridge, Massachusetts: Blackwell.

Ozaki, T. 1979. 'Maximum likelihood estimation of Hawkes' self-exciting point processes'. *Annals of the Institute of Statistical Mathematics* 31: 145–155.

Parlour, C. A., and D. J. Seppi 2008. 'Limit order markets: A survey'. In *Handbook of Financial Intermediation and Banking*, edited by Anjan V. Thakor, and Arnoud W. A. Boot, 63-95. Amsterdam: North-Holland.

Politi, M., and E. Scalas. 2008. 'Fitting the empirical distribution of intertrade durations'. *Physica A: Statistical Mechanics and its Applications* 387: 2025–2034.

Potters, M., and J. P. Bouchaud. 2003. 'More statistical properties of order books and price impact'. *Physica A: Statistical Mechanics and its Applications* 324: 133–140.

Preis, T., S. Golke, W. Paul, and J. J. Schneider. 2006. 'Multi-agent-based Order Book Model of financial markets'. *Europhysics Letters* 75: 510–516.

———. 2007. 'Statistical analysis of financial returns for a multiagent order book model of asset trading'. *Physical Review E* 76: 016108.

Privault, N. 2013. *Understanding Markov Chains*. Singapore: Springer.

Raberto, M., S. Cincotti, S. M. Focardi, and M. Marchesi. 2001. 'Agent-based simulation of a financial market'. *Physica A: Statistical Mechanics and its Applications* 299: 319–327.

Riley, J. D. 1955. 'Solving Systems of Linear Equations with a Positive Definite, Symmetric, but possibly Ill-conditioned Matrix'. *Mathematical Tables and Other Aids to Computation* 9: 96–101.

Seabom, J. B. 1991. *Hypergeometric Functions and their Applications.* New York: Springer.

Seber, G. A. F., and A. J. Lee. 2003. *Linear Regression Analysis.* Hoboken: Wiley.

Silva, A. C, and V. M. Yakovenko. 2007. 'Stochastic volatility of financial markets as the fluctuating rate of trading: An empirical study'. *Physica A: Statistical Mechanics and its Applications* 382: 278–285.

Slanina, F. 2001. 'Mean-field approximation for a limit order driven market model'. *Physical Review E* 64: 056136.

Slanina, F. 2008. 'Critical comparison of several order-book models for stock-market fluctuations'. *The European Physical Journal B* 61: 225–240.

Smith, E., J. D. Farmer, L. Gillemot, and S. Krishnamurthy. 2003. 'Statistical Theory of the Continuous Double Auction'. *Quantitative Finance* 3: 481–514.

Stigler, G. J. 1964. 'Public Regulation of the Securities Markets'. *Journal of Business* 37: 117–142.

Tibshirani, R. 1996. 'Regression Shrinkage and Selection via the Lasso'. *Journal of the Royal Statistical Society. Series B* 58: 267–288.

Tóth, B., Y. Lempériére, C. Deremble, J. de Lataillade, J. Kockelkoren, and J. P. Bouchaud. 2011. 'Anomalous price impact and the critical nature of liquidity in financial markets'. *Physical Review X* 1: 021006.

Vidyamurthy, G. 2004. *Pairs Trading: Quantitative Methods and Analysis.* Hoboken: Wiley.

Whitt, W. 2007. 'Proofs of the martingale FCLT'. *Probability Surveys* 4: 268–302.

Wyart, M., and J. P. Bouchaud. 2007. 'Self-referential behavior, overreaction and conventions in financial markets'. *Journal of Economic Behavior & Organization* 63: 1–24.

Zheng, B., E. Moulines, and F. Abergel. 2012. 'Price Jump Prediction in Limit Order Book'. *Journal of Mathematical Finance* 3: 242–255.

Zheng, B., F. Roueff, and F. Abergel. 2014. 'Modelling Bid and Ask Prices Using Constrained Hawkes Processes: Ergodicity and Scaling Limit'. *SIAM Journal on Financial Mathematics* 5: 99–136.

Zou, H., and T. Hastie. 2005. 'Regularization and Variable Selection via the Elastic Net'. *Journal of the Royal Statistical Society* 67: 301–320.